Probability Through Data: Interpreting Results from Frequency Tables

TEACHER'S EDITION

DATA-DRIVEN MATHEMATICS

Patrick W. Hopfensperger, Henry Kranendonk, and Richard Scheaffer

Dale Seymour Publications®
White Plains, New York

This material was produced as a part of the American Statistical Association's Project "A Data-Driven Curriculum Strand for High School" with funding through the National Science Foundation, Grant #MDR-9054648. Any opinions, findings, conclusions, or recommendations expressed in this publication are those of the authors and do not necessarily reflect the views of the National Science Foundation.

Managing Editor: Alan MacDonell

Senior Mathematics Editor: Nancy R. Anderson

Consulting Editor: Maureen Laude

Production/Manufacturing Director: Janet Yearian

Production/Manufacturing Manager: Karen Edmonds

Production Coordinator: Roxanne Knoll

Design Manager: Jeff Kelly

Text and Cover Design: Christy Butterfield

Cover Photo: Steve Satushek, Image Bank

This book is published by Dale Seymour Publications®, an imprint of Addison Wesley Longman, Inc.

Dale Seymour Publications
10 Bank Street
White Plains, NY 10602
Customer Service: 800-872-1100

Printed in the United States of America.

Order number DS21171

ISBN 1-57232-226-8

2 3 4 5 6 7 8 9 10-ML-03 02 01 00

This Book Is Printed
On Recycled Paper

Authors

Patrick W. Hopfensperger
Homestead High School
Mequon, Wisconsin

Henry Kranendonk
Rufus King High School
Milwaukee, Wisconsin

Richard Scheaffer
University of Florida
Gainesville, Florida

Consultants

Jack Burrill
National Center for Mathematics
Sciences Education
University of Wisconsin-Madison
Madison, Wisconsin

Emily Errthum
Homestead High School
Mequon, Wisconsin

Maria Mastromatteo
Brown Middle School
Ravenna, Ohio

Vince O'Connor
Milwaukee Public Schools
Milwaukee, Wisconsin

Jeffrey Witmer
Oberlin College
Oberlin, Ohio

Data-Driven Mathematics Leadership Team

Gail F. Burrill
National Center for Mathematics
Sciences Education
University of Wisconsin-Madison
Madison, Wisconsin

Miriam Clifford
Nicolet High School
Glendale, Wisconsin

James M. Landwehr
Bell Laboratories
Lucent Technologies
Murray Hill, New Jersey

Richard Scheaffer
University of Florida
Gainesville, Florida

Kenneth Sherrick
Berlin High School
Berlin, Connecticut

Acknowledgments

The authors thank the following people for their assistance
during the preparation of this module:

- The many teachers who reviewed drafts and participated in the field
  tests of the manuscripts

- The members of the *Data-Driven Mathematics* leadership team, the
  consultants, and the writers

- Kathryn Rowe and Wayne Jones for their help in organizing the field-
  test process and the Leadership Workshops

- Jean Moon for her advice on how to improve the field-test process

- The many students and teachers from Rufus King High School and
  Homestead High School who helped shape the ideas as they were
  being developed

Table of Contents

About *Data-Driven Mathematics*

Historically, the purposes of secondary-school mathematics have been to provide students with opportunities to acquire the mathematical knowledge needed for daily life and effective citizenship, to prepare students for the workforce, and to prepare students for postsecondary education. In order to accomplish these purposes today, students must be able to analyze, interpret, and communicate information from data.

Data-Driven Mathematics is a series of modules meant to complement a mathematics curriculum in the process of reform. The modules offer materials that integrate data analysis with high-school mathematics courses. Using these materials helps teachers motivate, develop, and reinforce concepts taught in current texts. The materials incorporate the major concepts from data analysis to provide realistic situations for the development of mathematical knowledge and realistic opportunities for practice. The extensive use of real data provides opportunities for students to engage in meaningful mathematics. The use of real-world examples increases student motivation and provides opportunities to apply the mathematics taught in secondary school.

The project, funded by the National Science Foundation, included writing and field testing the modules, and holding conferences for teachers to introduce them to the materials and to seek their input on the form and direction of the modules. The modules are the result of a collaboration between statisticians and teachers who have agreed on the statistical concepts most important for students to know and the relationship of these concepts to the secondary mathematics curriculum.

A diagram of the modules and possible relationships to the curriculum is on the back cover of each Teacher's Edition of the modules.

Using This Module

Why the Content Is Important

A student's introduction to probability is typically connected to a study of percent or an overview of statistics. Although the general idea of probability is not difficult, students frequently view this special branch of mathematics as a complex study of game theory and counting techniques. An introduction to probability is rarely able to provide students with an adequate range of the questions investigated by this topic. Probability is certainly about counting. It is also, however, about ordering, organizing, and summarizing. This module develops this topic by involving students in an active process of forming a collection of data. The formal counting techniques are not important at this time. This module further develops examples in which probability is based on organizing outcomes. The range of questions presented to students in this module expands the type of applications traditionally studied in an introduction to probability.

The content of this module involves students in creating data sets and in observing collected data. In each case, students are required to organize the data to form frequency charts and relative frequency charts. This information is then generalized and used to develop the introductory topics of probability. Symbolic representation of the process is primarily an attempt to organize and summarize the main topics.

Similar to other introductory books related to probability, this module works with problems involving tossing a coin or landing in a certain area of a spinner. The connections to games, however, are minimal. The initial goal is to develop the observations from relatively simple simulations as a process to trace the beginning roots of what is probability and the types of questions associated with this topic. After the initial topics are introduced, students are directed at organizing collected data into charts and graphs. This information is used to introduce new questions and new problems. The initial foundation of probability is still there, but the range of problems and situations related to the topic is expanded.

This module develops data sets from a variety of applications. In the third unit of the module, *Data Tables and Probability,* the primary format for organizing the information is two-way tables. Students are guided to organize frequency charts, relative frequency charts, and graphs from a two-way table. Questions introducing students to the topics of compound events, complementary events, and conditional probabilities are developed. This relatively simple method of organizing data provides an excellent opportunity to expand the questions involving probability. The topics developed in Unit III also guide students into an introduction of *association* in Unit IV.

Association is not a simple idea. This module, however, provides an initial definition of association related to data collected from surveys and research experiments. Although students are not directed at measuring association through a formal statistic, this module provides them a clearer idea of analyzing this complex topic by relating it to their work with probability. From a statistical perspective, students work with data to summarize the answer to a question or the solution to a problem. Unlike other problems, however, the students are guided into realizing that the summary of their initial questions leads them into more complex questions. This is intentional, as this module attempts to use data from research studies as an example of real-world applications.

Mathematics Content

Students will be able to

- Organize collected data into charts.
- Summarize data into percents.
- Identify and interpret relative frequencies as probability estimates.
- Graph frequency data into bar and line graphs.
- Create conditional relative frequency charts.
- Interpret conditional relative frequencies as an estimate of probability.
- Graph information from a chart and interpret the graph as a measure of association.

Statistics Content

Students will be able to

- Determine relative frequencies from a simulated collection of data.
- Find expected values from relative frequencies.
- Find and interpret expected values as an estimate of an outcome from a specific problem.
- Interpret graphical summaries as an indication of association of two variables.
- Analyze survey data presented in two-way tables.

Instructional Model

The instructional emphasis in *Probability Through Data,* as in all of the modules in *Data-Driven Mathematics,* is on discourse and student involvement. Each lesson is designed around a problem or mathematical situation and begins with a series of introductory questions or scenarios that can prompt discussion and raise issues about that problem. These questions can involve students in thinking about the problem and help them understand why such a problem might be of interest to someone in the world outside the classroom. The questions can be used in whole-class discussion or in student groups. In some cases, the questions are appropriate to assign as homework to be done with input from families or from others not a part of the school environment.

These opening questions are followed by discussion issues that clarify the initial questions and begin to shape the direction of the lesson. Once the stage has been set for the problem, students begin to investigate the situation mathematically. As students work their way through the investigations, it is important that they have the opportunity to share their thinking with others and to discuss their solutions in small groups and with the entire class. Many of the exercises are designed for groups in which each member does one part of the problem and the results are compiled for final analysis and solution. Multiple solutions and solution strategies are also possible, and it is important for students to recognize these situations and to discuss the reasoning behind different approaches. This will provide each student with a wide variety of ways to build his or her own understanding of the mathematics.

In many cases, students are expected to construct their own understanding by thinking about the problem from several perspectives. They do need, however, validation of their thinking and confirmation that they are on the right track, which is why discourse among students, and between students and teacher, is critical. In addition, an important part of the teacher's role is to help students link the ideas within an investigation and to provide an overview of the "big picture" of the mathematics within the investigation. To facilitate this, a review of the mathematics appears in the summary following each investigation.

Each investigation is followed by a Practice and Applications section in which students can revisit ideas presented within the lesson. These exercises may be assigned as homework, given as group work during class, or omitted altogether if students are ready to move ahead.

Periodically, student assessments occur in the student book. These can be assigned as long-range take-home tasks, as group assessment activities, or as in-class work. The assessment pages provide a summary of the lessons up to that point and can serve as a vehicle for students to demonstrate what they know and what they can do with the mathematics. Commenting on the strategies students use to solve a problem can encourage students to apply different strategies. Students also learn to recognize those strategies that enable them to find solutions efficiently.

Where to Use the Module in the Curriculum

This module is an introduction to the topics of probability. It is appropriate to use this module in an introduction to algebra course or a first-year algebra course. The lessons would also work well in an integrated course in which the topics of probability are introduced. Students are not required to have had any previous work with probability. Although this module presents the basic topics of probability, it covers several topics that are also challenging and appropriate in a more advanced course. The introductory nature of this module provides the background for various applications of probability.

This module is an appropriate starting point for many topics that are treated with more rigor and application in an advanced algebra module entitled *Probability Models,* another module in the *Data-Driven Mathematics* series.

Pacing/Planning Guide

The table below provides a possible sequence and pacing of the lessons in
this module:

LESSON	OBJECTIVES	PACING
Unit I: Probability as Relative Frequency		
Lesson 1: Relative Frequency	Organize data collected from a probability experiment into a table; convert outcomes from a probability experiment into relative frequencies; recognize that increasing the number of trials will cause the experimental probability to approach the theoretical probabilty.	1–2 class periods
Lesson 2: Applying Relative Frequency	Estimate the theoretical probability of an event based on the convergence of the relative frequencies.	1–2 class periods
Unit II: Simulation		
Lesson 3: Designing a Simulation	Recognize probability problems that are of the form "How many successes occur in *n* repetitions of an event?"; simulate the distribution of the number of successes in *n* repititions of an event; use simulated distributions to make decisions.	1 class period
Lesson 4: Waiting for Success	Recognize probability problems that are of the form "How long will I have to wait until the first success occurs?"; simulate the distribution of the number of trials until a success is achieved; use simulated distributions to make decisions.	1 class period
Assessment for Units I and II	Apply the concepts of simulation to answer probability questions.	1 class period
Unit III: Data Tables and Probability		
Lesson 5: Probability and Survey Results	Find an estimate of the probability of an event given the results of a survey.	2 class periods
Lesson 6: Compound Events	Determine the probability that both event *A* and event *B* will occur; determine the probability that either event *A* or event *B* will occur; organize data on two variables in a two-way table.	2 class periods
Lesson 7: Complementary Events	Find the probability of the complement of an event.	1/2 class period
Lesson 8: Conditional Probability	Construct and interpret relative frequencies from columns or from rows of a table; interpret column or row relative frequencies as conditional probabilities.	1-1/2 class periods
Assessment for Unit III	Apply the concepts of compound events, complementary events, and conditional probability.	1 class period

Unit IV: Understanding Association		
Lesson 9: Association	Understand how conditional probability can be used to measure association between two variables.	1 class period
Lesson 10: Constructing Tables from Conditional Probabilities	Determine expected frequencies from conditional probabilities; interpret probability statements by constructing an appropriate table of expected frequencies; determine unconditional probabilities from a table of expected frequencies.	1 class period
Lesson 11: Comparing Observed and Expected Values	Simulate the variability that may be attached to table of expected values; make decisions about the presence of association based on this variability.	2 class periods
Assessment for Unit IV	Apply the concepts of conditional probability and simulation to decide if there is an association between two variables.	1 class period
Project: Analyzing Survey Results	Analyze results of your own survey.	2-1/2–3-1/2 class periods
		approximately 4 weeks total time

Use of Teacher Resources

At the back of this Teacher's Edition are the following:

- Quizzes for selected lessons
- End-of-Module Test
- Solution Key for quizzes and test
- Activity Sheets

These items are referenced in the *Materials* section at the beginning of the lesson commentary.

LESSON	RESOURCE MATERIALS
Unit I: Probability as Relative Frequency	
Lesson 1: Relative Frequency	• *Activity Sheet 1* (Problem 2) • *Activity Sheet 2* (Problem 7) • *Activity Sheet 3* (Problems 11 and 14)
Lesson 2: Applying Relative Frequency	• *Activity Sheet 4* (Problem 1) • *Activity Sheet 5* (Problem 1) • *Activity Sheet 6* (Problems 1 and 5) • *Activity Sheet 7* (Problem 3) • *Activity Sheet 8* (Problem 7) • Lessons 1 and 2 Quiz
Unit II: Simulation	
Lesson 3: Designing a Simulation	
Lesson 4: Waiting for Success	• Lessons 3 and 4 Quiz
Assessment for Units I and II	• *Activity Sheet 9* (Problem 1) • *Activity Sheet 10* (Problem 1)
Unit III: Data Tables and Probability	
Lesson 5: Probability and Survey Results	• *Activity Sheet 11* (Problem 1) • *Activity Sheet 12* (Problem 2)
Lesson 6: Compound Events	• *Activity Sheet 13* (Problems 5, 6, 7, 15, and 16)
Lesson 7: Complementary Events	• Lessons 6 and 7 Quiz
Lesson 8: Conditional Probability	
Assessment for Unit III	
Unit IV: Understanding Association	
Lesson 9: Association	

LESSON	RESOURCE MATERIALS
Lesson 10: Constructing Tables from Conditional Probabilities	• Lessons 9 and 10 Quiz
Lesson 11: Comparing Observed and Expected Values	
Assessment for Unit IV	• End-of-Module Test
Project: Analyzing Survey Results	

Where to Use the Module in the Curriculum

This module is an introduction to the topics of probability. It is appropriate to use this module in an introduction to algebra course or a first-year algebra course. The lessons would also work well in an integrated course in which the topics of probability are introduced. Students are not required to have had any previous work with probability. Although this module presents the basic topics of probability, it covers several topics that are also challenging and appropriate in a more advanced course. The introductory nature of this module provides the background for various applications of probability.

This module is an appropriate starting point for many topics that are treated with more rigor and application in an advanced algebra module entitled *Probability Models,* another module in the *Data-Driven Mathematics* series.

Technology

The lessons extensively involve calculations requiring the use of a scientific calculator. In several lessons, simulations are developed in which students could use a random-number table, however, a more effective process would be to generate the values through a special option on most graphing calculators. Descriptions of generating random numbers are highlighted in this Teacher's Edition using a TI-83 as an example. In some of the lessons, the data tables students are requested to create could easily be generated as lists in a graphing calculator. Explanations are provided in this Teacher's Edition for this possibility.

Several lessons require the students to develop bar graphs or line graphs. A spreadsheet application could be used to generate these graphs and would add to the depth of the applications. The graphs, however, could be developed by hand as they are basic applications of graphing concepts.

Probability as Relative Frequency

LESSON 1

Relative Frequency

Materials: coins or two-sided counters, dice, *Activity Sheets 1–3*
Technology: graphing calculator with list capabilities or spreadsheet program (optional)
Pacing: 1–2 class periods

Overview

In this lesson, students use data collected from tossing a coin to investigate the concept of relative frequency. Students collect data on 30 coin tosses and record the data in different ways. First, the students record the data as *Heads* or *Tails*. The students then record the number of heads observed in 30 tosses and combine their data with that of others in the class. The class results are then graphed and converted to relative frequency. The understanding of relative frequency is crucial for the rest of the module. Students are then asked to investigate how the relative frequency of heads changes as the number of tosses increases by finding the cumulative number of heads and tails and converting these values into relative frequency. The last part of the lesson asks students to construct a line graph of the relative frequency.

Teaching Notes

Questions that center around students' notions of probability can be used to start the lesson. Have students place on a scale from 0 to 1 how likely or unlikely it is that certain events will occur. As the lesson progresses, it is important to allow students the time to toss the coins and record the outcomes. Using their own data helps students understand the concept of cumulative relative frequency. It is important for students to observe that as the number of tosses increases, the relative frequency of heads will approach 0.5. Students already know the probability of a head is 0.5, but this lesson demonstrates the use of a line graph in visualizing the concept of probability.

Students could use a spreadsheet program or a graphing calculator to collect their data. Students could enter a 0 for a tail and a 1 for a head. Following is an example showing how a graphing calculator could be used to calculate relative frequencies.

L1	L2	L3	L4
1	0	0	0
2	0	0	0
3	1	1	.3333
4	0	1	.25
5	1	2	.4

L1 = toss number
L2 = 0 for tail and 1 for head
L3 = cumsum(L2)
L4 = L3/L1

A graphing calculator could also be used to draw a line graph. For the example above, students would graph L1 as the X list and L4 as the Y list.

At the end of the lesson, students should recognize that as the number of trials increases, the line graph approaches the theoretical probability of an event.

Solution Key

Discussion and Practice

1. a. Most students will answer this question with 15 heads. Variations around the value of 15 indicate students anticipate one-half of the tosses will be heads.

   b. Outcomes will vary.

STUDENT PAGE 3

LESSON 1

Relative Frequency

If the weather forecaster said that there is 60% chance of rain today, would you carry an umbrella?

What if the weather forecaster said 40%?

Weather forecasting depends on a forecaster's knowledge of probability and analysis of data collected over a period of time.

The theory of probability is a part of mathematics that deals with uncertainty. The foundations for the theory were laid in the 1500s and 1600s by mathematicians who were interested in questions about games of chance. Since that time, probability has been associated with many other fields, such as meteorology.

INVESTIGATE
Coin Tossing

If you toss a coin into the air, will it land heads or tails? Sometimes it lands heads, sometimes tails. You cannot say for sure what the next outcome will be. The tossing of a coin is an example of a *random event*. For a random event, individual outcomes are unpredictable; but after a great many trials, a long-term pattern may emerge. In this lesson, you will investigate the long-term patterns in the number of heads that occur after tossing a coin many times.

Discussion and Practice

1. Consider tosses of a coin.

   a. Predict the number of heads that will occur when a coin is tossed 30 times. Explain your prediction.

   b. Actually toss a coin 30 times and record the outcomes in a table similar to the one on page 4.

OBJECTIVES

Organize data collected from a probability experiment into a table.

Convert outcomes from a probability experiment into relative frequencies.

Recognize that increasing the number of trials will cause the experimental probability to approach the theoretical probability.

c. Outcomes will vary. Anticipate answers of approximately 15 heads and 15 tails; however, relatively few students in a class will have exactly this result as explained in the following problems.

d. Students are directed to compare their results from tossing the coin in part c to the predicted outcome in part a. Answers will vary.

2. This problem provides data for working with the main ideas of this lesson. Students were directed to toss the coin 30 times. Discuss why the outcomes ranging from 0 heads to 30 heads should be included on their summary table. It is very unlikely outcomes of 0, 1, 2, or 28, 29, 30 will occur. If appropriate, ask students why they think that those results are unlikely.

STUDENT PAGE 4

Toss Number	Outcome (H or T)	Toss Number	Outcome (H or T)
1	———	16	———
2	———	17	———
3	———	18	———
.	———	.	———
.	———	.	———
.	———	.	———
15	———	30	———

c. Record the total number of heads and tails.

Outcome	Number
H	———
T	———

d. Compare the number of heads that you observed from the experiment with the number you predicted.

2. Ask the other students in class how many heads each one observed in his or her 30 tosses of the coin. In a table similar to the one shown below or *Activity Sheet 1*, record the results from all the students in your class.

Number of Heads	Tally	Frequency
0	———	———
1	———	———
2	———	———
3	———	———
4	———	———
5	———	———
6	———	———
.	———	———
.	———	———
.	———	———
30	———	———

3. The first part of this problem directs students to develop a line plot of the data collected in Problem 2. The directions outlined in the problem indicate a simple histogram can be constructed by simply aligning an X over the outcomes for each student. A histogram can also be developed with students using a graphing calculator. The format for developing a histogram varies with the type of calculator; however, the following steps indicate the process for creating a histogram with a TI-83.

Step 1: Clear two lists (for example, L1 and L2). The first list (L1) will be used to indicate the number of heads, and the second list (L2) will be used to record the frequency of the specific outcome.

Step 2: Record the values indicated in each of the lists.

L1	L2
0	0
1	0
2	0
•	•
•	•
12	3
13	1
14	4
15	4
16	7
•	•
•	•
•	•
29	0
30	0

Step 3: Select **2ND STAT PLOT**. Set the setting of one of the plot options to **On.**

Step 4: Select the histogram option under **Type.**

Step 5: For **Xlist,** enter **L1** and for **Freq** indicate **L2.**

Step 6: Select **WINDOW** and adjust the values of the *x* and *y* variables to appropriately view the

STUDENT PAGE 5

3. Use the data collected from each class member to construct a graph of the frequencies of the number of heads from 30 tosses by writing an X above the number of heads for each time that number occurred in class. For example, if six students counted 13 heads when they tossed the coin 30 times, you would write six Xs above the number 13, as shown below.

Class Results for Number of Heads

```
        X
        X
        X
        X
        X
        X
  0 1 2 3 4 5 6 7 8 9 10 11 12 13 14 15 16 17 18 19 20 21 22 23 24 25 26 27 28 29 30
                                Number of Heads
```

a. Describe the shape of the graph of the class data.

b. Study the graph. Which number of heads occurred most often? What percent of all the tosses in class does this value represent?

c. If you were to toss another coin 30 times, would you be surprised if you observed 12 heads? 18 heads? 24 heads? Explain how you decided.

The ratio of heads to the total number of tosses is called the **relative frequency** of heads. This relative frequency can be expressed as a fraction, a decimal, or a percent.

$$\text{relative frequency of heads} = \frac{\text{number of heads tossed}}{\text{total number of tosses}}$$

4. Consider tosses of a coin.

a. What would you expect the relative frequency of heads to equal after 30 tosses of a coin? Compare this value with the results from your 30 tosses of the coin.

b. If a coin is tossed 1000 times, what value would you predict for the relative frequency of heads?

c. If the relative frequency of heads for 50 tosses of a coin is 0.54, how many heads were observed?

---

data set. For the above example, the following values were selected:

WINDOW
    Xmin = 0
    Xmax = 30
    Xscl = 1
    Ymin = 0
    Ymax = 10
    Yscl = 1
    Xres = 1

Step 7: Assuming all functions are clear, select **GRAPH.**

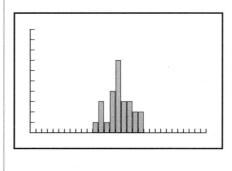

STUDENT PAGE 6

a. Students should observe the increase in the frequency of outcomes. Although the specific value which occurs the most often depends on the data set, it is anticipated that the most frequent value will occur around 15 heads. Interesting discussions result in students' observing that 15 is often not the most frequent outcome.

b. Results will vary depending on the data set. Students are to record the number of heads that occurred the most often and the percent of this occurrence out of the total number of students involved in the class.

c. Any of the outcomes of 12, 18, and 24 could occur; however, the likelihood of any of these outcomes is low. 24 heads would be the most unlikely, as its distance from the expected outcomes of 14, 15, or 16 is the greatest.

4. a. Answers will vary; however, it is anticipated that students will estimate the relative frequency to be close to 0.5, or 50%.

b. Here, again, it is anticipated that students will estimate the relative frequency to be 50%. (Make sure the students are estimating the *relative* frequency and not the *actual* frequency. If their estimate of the number of heads is 500, then the relative frequency is 0.5, or 50%.)

c. $50 \times 0.54 = 27$; therefore, 27 heads were observed.

d. $80 \times 0.55 = 44$; therefore, 44 heads occurred. This would indicate that 36 tails occurred.

5. Answers will vary depending on a student's data set. It is anticipated that the relative frequency for heads and the relative frequency for tails would be similar.

d. If the relative frequency of heads for 80 tosses of a coin is 0.55, what is the relative frequency of tails?

5. Calculate the relative frequency for both heads and tails for your 30 tosses. Express your answer as a fraction, a decimal, and a percent.

6. Use the data gathered by the class.

   a. What is the total number of heads and tails for your entire class?

   b. Calculate the relative frequency of heads and tails for the class data. Express your answer as a fraction, a decimal, and a percent.

   c. Compare the value of the relative frequency of heads from part b with the value of the relative frequency of heads calculated from 30 tosses of a coin in Problem 5.

The table below shows the results of 10 tosses of a coin. This table contains new columns, *Cumulative Number of Heads* and *Cumulative Number of Tails*. The values reflect the total number of heads or tails after each toss. For example, on the third toss a head was observed for the first time; so the cumulative number of heads is 1. The fourth toss was a tail; therefore, the cumulative number of heads remained 1.

Relative Frequency of Heads

Toss Number	Outcome	Cumulative Number of Heads	Relative Frequency of Heads	Cumulative Number of Tails	Relative Frequency of Tails
1	T	0	$\frac{0}{1} = 0$	1	$\frac{1}{1} = 1.00$
2	T	0	$\frac{0}{2} = 0$	2	$\frac{2}{2} = 1.00$
3	H	1	$\frac{1}{3} \approx 0.33$	2	$\frac{2}{3} \approx 0.66$
4	T	1	$\frac{1}{4} = 0.25$	3	$\frac{3}{4} = 0.75$
5	H	2	$\frac{2}{5} = 0.4$	3	$\frac{3}{5} = 0.60$
6	H	3	$\frac{3}{6} = 0.5$	3	$\frac{3}{6} = 0.5$
7	H	4	$\frac{4}{7} \approx 0.57$	3	$\frac{3}{7} \approx 0.43$
8	T	4	$\frac{4}{8} = 0.5$	4	$\frac{4}{8} = 0.5$
9	T	4	$\frac{4}{9} \approx 0.44$	5	$\frac{5}{9} \approx 0.56$
10	T	4	$\frac{4}{10} = 0.4$	6	$\frac{6}{10} = 0.6$

6. a. Answers will vary depending on the class data. The total number of tosses will significantly increase as each student tosses the coin 30 times. Therefore, the number of students multiplied by 30 will represent the total number of tosses. The number of heads and the number of tails will increase correspondingly.

b. Answers will vary depending on the combined data from students in the class.

c. Answers will vary. It is anticipated, however, that the relative frequencies will be similar.

STUDENT PAGE 7

7. Students return to the 30 tosses they recorded earlier. Completing a table as directed in this problem involves many calculations. Use of a calculator is critical. The columns to watch are the "Cumulative Number of Heads" and the "Cumulative Number of Tails." The relative frequencies are based on comparing the cumulative numbers to the number of tosses.

8. **a.** The cumulative number of heads plus the cumulative of tails for any row will equal the number of tosses for that row.

   b. The sum of the relative frequencies will equal 1.00, or 100%.

9. Students will likely observe that the relative frequencies are erratic in the beginning. As the total number of tosses is a small number in the beginning of the chart, the relative frequencies noticeably change from one toss to the next. The relative frequencies demonstrate smaller changes or less erratic changes as the total number of tosses increase. It is also anticipated that students will observe the relative frequencies approach the predicted value of 0.50, or 50%.

As the number of tosses increases, the relative frequency of heads changes. This value is calculated by dividing the *cumulative number of heads* by the *toss number*. For example, after the fifth toss there were 2 heads, so the relative frequency is $\frac{2}{5}$, or 0.4.

7. Use the results from your 30 tosses to complete a table similar to the one shown below, or use the table on *Activity Sheet 2*. Give the relative frequency of heads and tails as a decimal.

Relative Frequency of Heads

Toss Number	Outcome	Cumulative Number of Heads	Relative Frequency of Heads	Cumulative Number of Tails	Relative Frequency of Tails
1	———	———	———	———	———
2	———	———	———	———	———
3	———	———	———	———	———
.	———	———	———	———	———
.	———	———	———	———	———
.	———	———	———	———	———
30	———	———	———	———	———

8. Study the table you just completed.

   a. What is the sum of the cumulative numbers of heads and tails for any row?

   b. What is the sum of the relative frequency of heads and the relative frequency of tails for any row of the table?

9. As the number of tosses increases, describe what happens to the relative frequency of heads.

10. 0.57, or 57%, after 7 tosses. This is based on the values displayed in the table in Problem 7. The relative frequency of heads for 7 tosses was $\frac{4}{7} \approx 0.57$.

11. Students can construct the line graph on a grid similar to the one presented with this problem. You might also encourage students to develop the line graph with a graphing calculator or spread-sheet. Here again, the format for developing a line graph depends on the specific graphing calculator used. The following steps describe the process for setting up a line graph of the relative frequencies of heads for the table in Problem 7 using a TI-83.

Step 1: Clear lists L1, L2, and L3.

Step 2: Enter number of tosses in L1, the cumulative number of heads in L2, and the relative frequency of heads in L3. The relative frequency values for L3 can be programmed by placing the cursor at the top of L3. Watch the bottom of the window and enter **L3 = L2/L1.**

L1	L2	L3
1	0	0
2	0	0
3	1	.33333
4	1	.25
5	2	.4
6	3	.5
7	4	.57143
8	4	.5
9	4	.44444
10	5	.5

Step 3: To develop a line graph of this data, select **2ND STAT PLOT.** Activate Plot 1 by turning it **On.**

Step 4: Select line graph from the **Type** selections. Enter L1 for the **Xlist** and L3 for the **Ylist.**

Step 5: Adjust the **WINDOW** settings so that this graph can be

STUDENT PAGE 8

10. The data you calculated in Problem 7 can also be displayed in a line graph of relative frequencies. The graph below is a line graph of the data in the table in Problem 6. Point *A* shows that after the third toss, the relative frequency of heads was approximately 0.33. What does point *B* show?

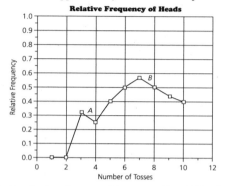

11. Use the data from your 30 tosses and a grid like the one below, also on *Activity Sheet 3.*

a. Construct a line graph of relative frequency of heads.

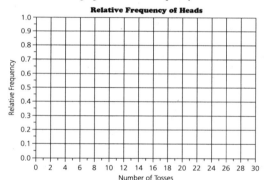

b. Describe any trends that you observe.

c. On the same graph, copy another student's line graph. Describe the similarities and differences in the graphs.

viewed. For this particular table, select:

Xmin = 0, **Xmax** = 10, **Xscl** = 1

Ymin = 0, **Ymax** = 1, **Yscl** = .1

Xres = 1

Or use **ZOOM STAT. ZOOM STAT** represents a convenient way to quickly set the window when graphing data lists. However, students should be guided in selecting window values by making appropriate decisions concerning the specific data sets. The **ZOOM STAT**

option should be used only after students understand the process and are capable of working through the details of graphing data sets.

Step 6: Select **GRAPH.**

As students become familiar with their own graphing calculator, their ability to obtain graphs for the problem sets involved in this module will improve. These first examples, although easily done without the calculator, are useful in introducing the calculator.

(11) a. Students again will observe the erratic changes in the relative frequency at the start of the graph. As the number of tosses increases, the changes in the relative frequency is less. In addition, the relative frequency is close to the expected value of 0.50, or 50%.

b. Answers will vary. It is anticipated that students will highlight the differences in the graphs. Find examples in which the beginning relative frequency is 0 for one student and 1.00 for another. It is anticipated that students will also highlight how the relative frequencies approach the expected value of 0.50, or 50%.

c. The horizontal line constructed at relative frequency 0.50 represents the expected value of the relative frequency of heads (or tails).

d. The primary change noted is that the erratic changes in the relative frequencies have lessened. As the number of tosses increases, the changes in the relative frequency are less pronounced and appear to "level off."

e. The relative frequency approaches the horizontal line constructed in part c.

12. Graph iii is the best estimate for the addition of 20 tosses. This problem provides opportunity to discuss one of the main objectives of the lesson. Although graph i approaches the expected outcome of 50%, the variation in the relative frequencies is too great for the number of tosses. If necessary, point out to the students how the jump from 0.5 to 0.75 for total tosses of 30 to 31 would not be possible. (In fact, the greatest possible increase in the relative

STUDENT PAGE 9

d. Draw a horizontal line across the graph at a relative frequency of 0.50. What does this line represent?

e. As the number of tosses increases, describe the changes in the line graph.

f. As the number of tosses increases, what relative frequency does the line graph seem to approach?

12. If you toss a coin 20 more times and combine these tosses with your previous 30 tosses, what do you think would happen to the line graph? Which of these three graphs comes closest to your expectation? Explain your answer.

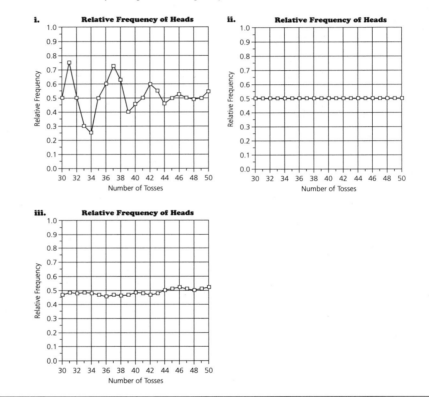

frequency would be from $\frac{15}{30}$, or 0.5, to $\frac{16}{31} \approx 0.52$).

Graph ii, although possible, is also unlikely because of the exact nature of the relative frequencies. Variation is going to occur. At this point, students should recognize the unlikelihood of exactly 0.5, or 50%, in their tosses.

13. The relative frequency of 0.5 or 50% indicates that the probability of observing a head in the toss of a coin is 50%. The greater the number of the tosses, the more the relative frequency is a good estimate of the probability.

Practice and Applications

14. a. The line graph developed by students will indicate whether they understand the objectives of this lesson. The relative frequency of even rolls will start at 0 or 1. As the number of rolls increases, the relative frequency will show less variation and converge to a value of 0.50, or 50%, as 3 of the 6 numbers are even. The following example highlights these observations.

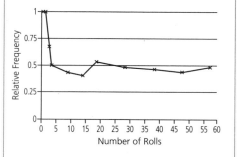

b. The line graph indicates the convergence of the relative frequency to the value of 0.50, or 50%. It also shows how the variation of the relative frequencies is more erratic in the beginning and eventually levels off to the expected value of 0.50. In this example, the relative frequency of 0.50 can be used to summarize the probability of getting an even number from a roll of the die.

c. Answers will vary. The properties summarized in this lesson should be observed in most of the examples.

STUDENT PAGE 10

The value that the line graph approaches can be thought of as the probability of observing a head in the toss of a coin. The *probability* of a head is the relative frequency of heads in a very great number of tosses.

13. What does a line graph of relative frequencies tell you about the probability of a head when tossing a coin?

Summary

This lesson was based on the results of 30 tosses of a coin. You observed that the relative frequency of heads varied with each toss, but this variability decreased as the number of tosses increased. After a great number of tosses, the relative frequency of heads approached the value 0.5, which is the probability of getting a head when tossing a coin.

Practice and Applications

14. A fair number cube has six faces that are numbered from 1 to 6. A student rolls the cube 60 times and records whether the face that turns up is an even or an odd number.

a. Make a sketch of a line graph that you think would show the results of the relative frequencies of getting an even number as the number of rolls of the die increases.

b. What does the line graph of relative frequencies tell you about the probability of getting an even number when rolling a fair die?

c. Work with your group, and roll a die 60 times. On *Activity Sheet 3*, construct a line graph of the relative frequencies of getting an even number. Compare your graph to the one in part a.

STUDENT PAGE 11

15. a. Consider the table below as a guideline.

b. Possible answer:

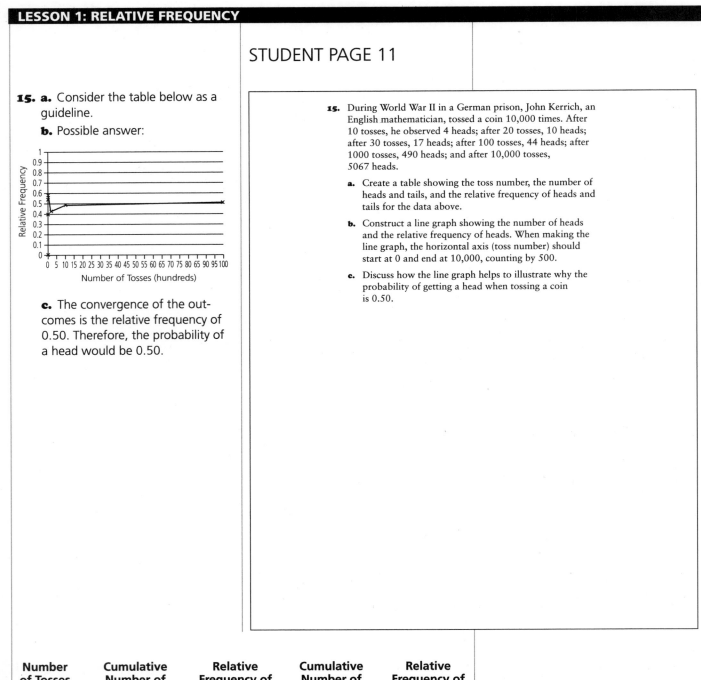

c. The convergence of the outcomes is the relative frequency of 0.50. Therefore, the probability of a head would be 0.50.

15. During World War II in a German prison, John Kerrich, an English mathematician, tossed a coin 10,000 times. After 10 tosses, he observed 4 heads; after 20 tosses, 10 heads; after 30 tosses, 17 heads; after 100 tosses, 44 heads; after 1000 tosses, 490 heads; and after 10,000 tosses, 5067 heads.

   a. Create a table showing the toss number, the number of heads and tails, and the relative frequency of heads and tails for the data above.

   b. Construct a line graph showing the number of heads and the relative frequency of heads. When making the line graph, the horizontal axis (toss number) should start at 0 and end at 10,000, counting by 500.

   c. Discuss how the line graph helps to illustrate why the probability of getting a head when tossing a coin is 0.50.

Number of Tosses	Cumulative Number of Heads	Relative Frequency of Heads	Cumulative Number of Tails	Relative Frequency of Tails
0	0	0	0	0
10	4	0.4	6	0.6
20	10	0.5	10	0.5
30	17	0.57	13	0.43
100	44	0.44	56	0.56
1000	490	0.49	510	0.51
10,000	5067	0.51	4933	0.49

LESSON 2

Applying Relative Frequency

Materials: paper clips, bags of Skittles®, *Activity Sheets 4–8,*
Lessons 1 and 2 Quiz (optional)
Technology: graphing calculator or spreadsheet program
(optional)
Pacing: 1–2 class periods

Overview

In Lesson 1, students investigated events in which the
theoretical probability was known. In this lesson, students experiment with events in which the theoretical
probability is not known. Students begin by making a
spinner using a paper clip and *Activity Sheet 4.* The
students tally the times that the paper clip lands in
area A and calculate the relative frequency of landing
in area A. Then they construct a line graph of these
data. From the graph, students should observe that
the line approaches the theoretical probability that
the spinner will land in area A. A second activity
involves using small bags of Skittles and finding the
probability of randomly selecting a purple Skittle.
According to the Mars Company, approximately 20%
of the Skittles are purple.

Teaching Notes

This lesson is very similar to Lesson 1, with difference
being that the theoretical probabilities are not known.
Students should be able to collect data from the spinner experiment, organize the data in a chart, and calculate the relative frequency. It is important that
students observe that as the number of trials increases, the variation in the relative frequency decreases
and approaches a theoretical value.

As in Lesson 1, a graphing calculator could be used to
display the collected data. Students could enter a 0
when the spinner lands in area B and a 1 when the
spinner lands in area A. In the Skittle experiment, students could enter a 0 for a nonpurple Skittle and a 1
for a purple Skittle.

Follow-Up

Students could construct their own spinners and have
other students determine the relative frequency of
landing in a certain section. Students could also use
M&Ms® and investigate the relative frequency of
each color.

Skittles and M&Ms are registered trademarks of M&M/Mars, a division of Mars, Inc.

STUDENT PAGE 12

Applying Relative Frequency

Did you ever watch the TV show *Wheel of Fortune*? A contestant spins the wheel that has numerous pie-shaped sections labeled with amounts of money.

If you watch the show long enough and tally which amounts of money come up, do you think that a pattern will emerge?

OBJECTIVE

Estimate the theoretical probability of an event based on the convergence of the relative frequencies.

The experiment that you performed in Lesson 1 required you to toss a coin 30 times and record the outcomes. Based on the results, you found the relative frequency of heads and constructed a line graph of these frequencies. From this line graph, you observed that the relative frequency of heads varied with each toss of the coin, but the relative frequency converged on the value 0.50.

Probability describes the uncertainty of any single event by summarizing what happens after many trials of an experiment. In this lesson, you will observe the outcomes of some experiments and, based on your observations, estimate the probability of an event.

INVESTIGATE

Spinning Around

You will need a paper clip and a copy of the figure on page 13, reproduced on *Activity Sheet 4*.

STUDENT PAGE 13

Solution Key

Discussion and Practice

1. **a.** At this point in the module, students can make only general estimates of this probability. Estimates should be clearly less than 0.50, or 50%. The anticipated response is approximately $\frac{1}{3} \approx 0.33$, or 33%. Theoretical probability is about 0.38.

b. Answers will vary. Caution: It is possible for students to manipulate the outcomes of a spin by a simple nudge or partial spin. Encourage students to record only those responses in which the paper clip makes at least 2 spins. This will not guarantee randomness, but it will improve the outcomes. Some of the goals of this lesson will be unclear if the spins do not produce a nearly random outcome.

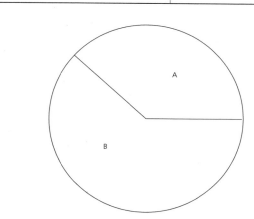

Discussion and Practice

1. Where is the spinner more likely to land, in the area marked A or the area marked B?

  a. Estimate the probability that on a random spin the spinner will land in area A.

  b. Use a paper clip on the tip of a pencil and the circle with areas A and B marked, as on *Activity Sheet 4*, to assemble a spinner. Then spin the spinner 30 times and record your results in a table similar to the one that follows. You many record your results in the first two columns of *Activity Sheet 5*.

Spin Number	Outcome (Area A or Area B)
1	_____
2	_____
3	_____
.	_____
.	_____
.	_____
30	_____

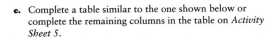

STUDENT PAGE 14

(1) c. Answers will vary according to the collected data.

d. As with the graphs in Lesson 1, the relative frequencies will level off or converge to a value. The specific value, however, is not as clear as in the previous lesson. It is anticipated that the relative frequencies will level off at a value between 0.33 and 0.43.

e. The changes in the relative frequencies are less obvious. In addition, the relative frequencies converge to a value between 0.33 and 0.43.

f. As indicated in the previous comments, the relative frequency levels off to a value between 0.33 and 0.43.

c. Complete a table similar to the one shown below or complete the remaining columns in the table on *Activity Sheet 5.*

Spin Number	Outcome (Area A or Area B)	Cumulative Number in Area A	Cumulative Number in Area B	Relative Frequency of Spins Landing in Area A
1	———	———	———	———
2	———	———	———	———
3	———	———	———	———
.	———	———	———	———
.	———	———	———	———
.	———	———	———	———
30	———	———	———	———

d. Use the data in the relative frequency column from the table you completed in part c to construct a line graph of the relative frequency of spins landing in area A for each spin. You may use the grid on *Activity Sheet 6.*

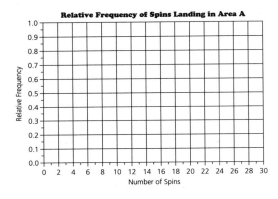

e. What happens to the line graph as the number of spins increases?

f. As the number of spins increases, what relative frequency does the line graph appear to approach?

STUDENT PAGE 15

g. Answers will vary. Similarities described by students should summarize the main ideas presented in Lesson 1. Differences are more apparent, as the actual values generated by each student will not be the same.

2. **a.** The line graph converges to a value representing the probability of landing in area A.

b. The specific value estimating the leveling off of the relative frequency will vary. As students compare their values, however, a similar range of values will emerge.

3. **a.** Answers will vary depending on the set of data collected. As with the observations in Lesson 1, the frequencies should increase to the expected outcome of landing in area A. Encourage students to use their graphing calculator and enter the data in two lists. For example, enter the numbers 0 through 30 in L1 and the corresponding frequencies in L2.

b. The graph should accurately represent the table values. The Xs provide a line plot of this data. If students recorded the values in part a as lists in a graphing calculator, then the calculator would be able to generate a histogram. An outline of the steps to accomplish this was developed in Lesson 1, Problem 3.

g. On the same graph, copy a classmate's graph. Describe the similarities and differences in the two graphs.

The relative frequency value the line graph approaches can be thought of as an estimate of the ***probability*** that the spinner will land in area A. Probability, as the value that the relative frequency approaches, can be generalized by the following rule:

As an experiment is repeated again and again, the relative frequency of occurrences of outcomes defining an event approaches the probability of the event. The value that the line graph of relative frequencies approaches can be used as an estimate for the ***probability of the event.***

2. Use your line graph from Problem 1d.

a. What does the line graph of relative frequency tell you about the probability of landing in area A on any spin?

b. If you were to spin 1000 times, what value would you predict for the relative frequency of the number of times landing in area A?

3. Ask other students in class for the number of times their spinner landed in area A.

a. Record the results in a table similar to the one shown below. You may use the table on *Activity Sheet 7.*

Number of Times Spinner Landed in Area A	Tally	Frequency
————	————	————
————	————	————
————	————	————
————	————	————
————	————	————
————	————	————

b. Use the data collected from each class member to construct a graph using a number line similar to the one that follows. Record the frequencies of the number of times the spinner landed in area A out of 30 spins by writing an X above the number of times for each time that number occurred in class.

(3) c. Answers will vary depending on the data generated by the class. It is anticipated that students will find outcomes between 10 and 13 as the most likely values for landing in area A.

d. It is unlikely that the number of times the spinner would land in A would be between 18 and 22 out of 30 spins. This represents an unusually high number of outcomes in the area unless students "fix" the spins. However, an outcome between 18 and 22, although unlikely, is nonetheless possible. (Observe that 18 to 22 is clearly more than half of the 30 attempts; yet area A represents less than half of the area of the circle.)

e. Answers will vary depending on the recorded values of the students. If an estimate of the relative frequency for landing in area A is 0.39, or 39%, then an estimate of the number of times the spinner might land in this area for 1000 spins is $0.39 \times 1000 = 390$.

As a result, an estimate of the number of times the spinner would land in this area is 390 times out of 1000. (Note: It is not likely, however, that exactly 390 would represent the number of times the spinner would land in this area.)

4. a. Answers will vary according to a student's estimate of the relative frequency. Assume 0.39, or 39%, is a student's estimate of the relative frequency. This student's estimate of the central angle, then, is 39% of 360, or $0.39 \times 360 = 140.4$ or approximately 140°.

b. The central angle defined by sector A is approximately 138°.

c. If the estimates used in parts a and b were used, then the angle calculated from the relative frequencies and the actual central

---

STUDENT PAGE 16

Class Results for Number of Times Spinner Landed in Area A

0 1 2 3 4 5 6 7 8 9 10 11 12 13 14 15 16 17 18 19 20 21 22 23 24 25 26 27 28 29 30
Number of Times

c. Read the graph to determine the most likely number of times the spinner would land in area A after 30 spins.

d. Do you think it would be likely for the spinner to stop from 18 to 22 times out of 30 spins in area A? Explain your answer.

e. If you were to spin the spinner 1000 times, how many times would you expect to land in the area marked A? How does this answer relate to the answer to Problem 2b?

4. Use the relative frequency that your line graph approached.

　a. Predict the measure of the central angle for area A.

　b. Use a protractor to find the measure of the angle.

　c. Compare the results from parts a and b.

Skittles

You will need a small bag of Skittles® for this problem.

5. Consider the question "What is the probability that a randomly chosen Skittle will be purple?"

　a. Open a small hole in the bag of Skittles. Allow only one Skittle at a time to fall out of the bag and note the color. Record the results in a table similar to the one that follows. As you collect the data, calculate the relative frequency of purple Skittles, and construct a line graph of the relative frequency. Use a grid like the one shown, reproduced on *Activity Sheet 6*.

Skittles is a registered trademark of M&M/Mars, a division of Mars, Inc.

---

angle of the sector would be very close.

5. a. Clearly this chart is similar to the previous charts used to summarize the relative frequencies of an event. In this case, the relative frequency of selecting a purple Skittle would be developed. This relative frequency is used to estimate the probability of selecting a purple Skittle from the bag of Skittles.

STUDENT PAGE 17

b. The number of Skittles selected by the student to estimate the probability of selecting a purple Skittle will vary. The estimate given by a student should be supported by a graph or a calculation of the relative frequencies. If a graph is used, the line graph of the relative frequencies should level off, or converge, at the estimate given by the student. This convergence is determined when little variation in the relative frequencies of selecting a purple Skittle is observed. A student may respond to this question by selecting all of the Skittles in the bag. In this way, the student is indicating that the probability of selecting a purple Skittle from this bag is the total number of purple Skittles divided by the total number of Skittles.

    i. Students' responses to this question should be supported by the statements indicated in part b.

    ii. A student's estimate of the probability of selecting a purple Skittle should be supported by the relative frequency of this event.

c. Comparing answers should highlight the differences and similarities of this experiment. Differences will result from actual differences in the number of purple Skittles in a bag. Students should be able to provide explanations based on their estimate of the relative frequencies obtained from this experiment.

d. A student's estimate of the relative frequency of selecting a purple Skittle should be multiplied by 65. This product is the student's estimate of the number of purple Skittles.

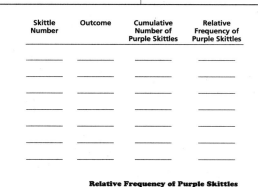

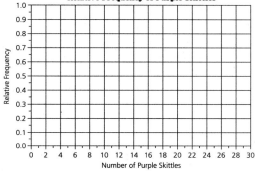

b. How many Skittles did you sample until you were confident that you knew the probability of randomly selecting a purple Skittle?

    i. Why did you stop at this number of Skittles?

    ii. What is your estimate for the probability of randomly selecting a purple Skittle?

c. Compare your estimate with those of other students in your class.

d. If there are 65 Skittles in a bag, how many purple Skittles do you expect to be in the bag? Explain your answer.

STUDENT PAGE 18

Practice and Applications

6. **a.** Answers will vary; however, the convergence of the relative frequencies should be in the range of 0.37 to 0.40.

   b. If an estimate of 0.38 were used for the relative frequency, then you would expect 0.38×500, or 190, successes to occur.

   c. The probability of success is based on the estimate of the value of the relative frequency as it levels off or converges. In this example, an estimate of 37% to 40% would represent the data displayed by the graph.

   d. Answers will vary. The situation might be the result of selecting a certain color from a jar of colored chips, or the situation of landing on a certain area of a spinner, or the situation of selecting football players from your school for an interview (assuming 38% of the students are football players).

7. **a.** Follow the directions as indicated.

   b. This problem requires students to collect the data by actually carrying out the problem. In addition, the students are expected to develop an organized table of the results, develop a line graph, and calculate the relative frequencies based on the topics introduced in this lesson.

Practice and Applications

6. The line graph below shows the relative frequency of a success for 30 trials of an experiment.

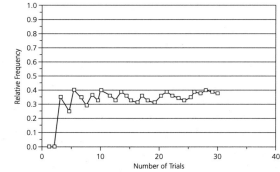

   a. As the number of trials increased, what relative frequency did the line graph approach?

   b. If this experiment were repeated for 500 trials, how many successes would you expect to see?

   c. What is your estimate for the probability of a success?

   d. Think of a situation that might be described by the graph.

7. For this problem, you will need scissors, tape, and a copy of the figure shown below, also on *Activity Sheet 8*.

   a. Cut out the figure. Then fold and tape it to form a square pyramid.

   b. Conduct an experiment to determine the probability that the pyramid will land on the square base when it is randomly tossed. Your work should include tables, graphs, and an explanation of how you arrived at an answer.

Simulation

LESSON 3

Designing a Simulation

Materials: random-number table (optional)
Technology: graphing calculator
Pacing: 1 class period

Overview

This lesson presents the basics of designing and conducting a simulation. The first investigation requires that students randomly select a number between 1 and 100. Students should use a graphing calculator or a random-number table. If using a random-number table, students select two digits and let 00 represent 100. After they have worked through the first simulation, it is important that students understand the steps that should be followed to conduct a simulation properly. The steps are listed in the summary of the lesson. The second simulation is taking a true-false test by guessing. The practice problems are designed to allow students an opportunity to practice setting up and conducting a simulation. For the multiple-choice questions, it is important that students know that the random numbers chosen to represent the correct answer must correspond to the theoretical probability of choosing the correct answer by chance.

Teaching Notes

In conducting the first simulation, students should collect their data in a table similar to those used in Lessons 1 and 2. An example of a table is given here.

Class data can also be collected. Each student should report how much he or she paid for the 60 sodas. A graph of these amounts should center around $45 (0.75×60).

Selection Number	Random Number	Amount Paid for Soda
1	_____	_____
2	_____	_____
•	_____	_____
•	_____	_____
•	_____	_____
60	_____	_____

In the second simulation, you could have students model taking a true-false test by having them write a random sequence of T and F and compare their lists to a random list of answers that you have generated. It is important to collect data from the class for the number of correct answers out of the five questions. The graph should center around two and three correct answers out of five. Each simulation that students conduct should contain each of the steps, with a written explanation of what was done and the results that were gathered.

Follow-Up

Students could investigate taking a test with a different number of choices. Students could also investigate how likely it is that they can pass a test by guessing.

STUDENT PAGE 21

Designing a Simulation

How likely is each situation described below?

> **A basketball player went to the free-throw line 10 times in a game and made all 10 shots.**
>
> **A student guessed all the answers on a 20-question true-false test and got 18 of them wrong.**
>
> **All the Skittles in a bag of Skittles were purple.**

INVESTIGATE

Unusual? Some people would say that the chances that these events would happen are very small. These three situations have certain common characteristics. They all involve repetition of the same event, and they all have as a goal counting the number of "successes" in a fixed number of repetitions.

Discussion and Practice

One Outcome per Trial

In this lesson, you will learn how to construct a simulation model for events of the type described above so that you can approximate their probabilities and decide for yourself whether or not the events are unusual.

1. Read the article on page 22 entitled *Non-Cents: Laws of Probability Could End Need for Change.*

OBJECTIVES

Recognize probability problems that are of the form "How many successes occur in n repetitions of an event?"

Simulate the distribution of the number of successes in n repetitions of an event.

Use simulated distributions to make decisions.

Solution Key

Discussion and Practice

1. a. Answers will vary. The article is written to promote the idea; therefore, it is anticipated that most students would indicate the system seems fair.

b. Answers will vary.

2. This problem is the first of several questions requiring students to determine a random number. This process can be handled in several ways. A simple random-number table could be used; however, students might be directed to generate random numbers using a graphing calculator or computer program. The TI graphing calculators can be used to generate random numbers within any given range. Although the process is quite simple, the components are located within several menu options. The following summary illustrates how to generate a random number between 1 and 100 using a TI-83. Modification of this process should be used for other calculators.

Step 1: Clear the home screen of the calculator. Paste the random integer option (**randInt**) into your home screen.

Select **MATH** and the **PRB** option from the displayed menu. The following menu options are now available.

```
MATH NUM CPX PRB

1:rand
2:nPr
3:nCr
4:!
5:randInt(
6:randNorm(
7:randBin(
```

STUDENT PAGE 22

NON-CENTS: LAWS OF PROBABILITY COULD END NEED FOR CHANGE

(*Milwaukee Journal*, May, 1992)

Chicago, Ill. [AP] Michael Rossides has a simple goal: to get rid of that change weighing down pockets and cluttering up purses.

And, he says, his scheme could help the economy.

"The change thing is the cutest aspect of it, but it's not the whole enchilada by any means," Rossides said.

His system, tested Thursday and Friday at Northwestern University in the north Chicago suburb of Evanston, uses the law of probability to round purchase amounts to the nearest dollar.

"I think it's rather ingenious," said John Deighton, an associate professor of marketing at the University of Chicago.

"It certainly simplifies the life of a businessperson and as long as there's no perceived cost to the consumer it's going to be adopted with relish," Deighton said.

Rossides' basic concept works like this:

A customer plunks down a jug of milk at the register and agrees to gamble on having the $1.89 be rounded down to $1 or up to $2.

Rossides' system weighs the odds so that over all transactions, the customer would end up paying an average $1.89 for the jug of milk but would not be inconvenienced by change.

That's where a random number generator comes in. With 89 cents the amount to be rounded, the amount is rounded up if the computerized generator produces a number from 1 to 89; from 90 to 100 the amount is rounded down.

Rossides, 29, says his system would cut out small transactions, reducing the cost of individual goods and using resources more efficiently.

The real question is whether people will accept it.

Rossides was delighted when more than 60% of the customers at a Northwestern business school coffee shop tried it Thursday.

Leo Hermacinski, a graduate student at Northwestern's Kellogg School of Management, gambled and won. He paid $1 for a cup of coffee and muffin that normally would have cost $1.30.

Rossides is seeking financial backing wants to test his patented system in convenience stores.

But a coffee shop manager said the system might not fare as well there.

"Virtually all of the clientele at Kellogg are educated in statistics, so the theories are readily grasped," said Craig Witt, also a graduate student. "If it were just to be applied cold to average convenience store customers, I don't know how it would be received."

a. Would you shop at a store that used the system described by Mr. Rossides?

b. Do you think that you would pay too much money under his proposal? Explain your reasoning.

Select **randInt** and **ENTER.** This will paste **randInt(** into your home screen.

```
randInt(
```

Step 2: At this point, enter the range of the desired random numbers. For selecting a random number from 1 to 100, enter the following.

```
randInt(1,100)
```

Select **ENTER** and a random number between 1 and 100 (including the possibility of 1 or 100) is displayed.

STUDENT PAGE 23

randInt(1,100)

46

To obtain more random numbers in the same range, simply press **ENTER** repeatedly.

If the above value of 46 is obtained by a student, then the student pays $1 for the soda. If the random number obtained is greater than 75, then the student pays nothing. Most of the random numbers generated would require the student to pay $1.

3. To generate 60 random numbers between 1 and 100, repeat the process described in step 2 and add the following parameter to the **randInt** function.

randInt(1,100,60)

Hit **ENTER** at this point, and 60 random integers between 1 and 100 are generated. To view the 60 values, use the arrow options.

randInt(1,100,60)

59 34 23 28 90 ...

a. Answers will vary depending on the random numbers generated. Expect 75% of the numbers to fall between 1 and 75, requiring the student to pay $1, and 25% of the numbers to fall between 76 and 100, requiring the student to pay nothing. If 45 of the numbers were less than or equal to 75, then the student would pay $45 for the 60 drinks. Direct students to compare

Suppose the soft-drink machine you use charges $0.75 per can. The scheme proposed by Mr. Rossides requires you to pay either $0 or $1, depending upon your selection of a random number. You select a random number between 1 and 100. If the number you select is 75 or less, you pay $1. If the number you select is greater than 75, then you pay nothing.

2. Simulate the buying of a can of soda by randomly selecting a number between 1 and 100 using a graphing calculator or a similar process. Did you have to pay $0 or $1 for the can of soda? Explain your answer.

3. The article suggests that things will even out in the long run. Suppose that over a period time, you purchase 60 drinks from this machine and use a random mechanism for payment each time. This can be simulated by choosing 60 random numbers between 1 and 100. Make such a selection of 60 random numbers using a calculator or a similar process and keep track of how much you paid for each soda.

  a. How many times did you pay $1? What is the total amount you paid for 60 drinks?

  b. If you had paid $0.75 for each drink, how much would you have paid for 60 drinks? Does the system proposed by Mr. Rossides seem to balance out in the long run?

4. Now suppose you are buying a box of cookies that cost $3.23. You pay either $3 or $4, depending on the outcome of a random-number selection.

  a. For what values of the random numbers should you pay $3? For what values of the random numbers should you pay $4?

  b. Use the rule you determined in part a to simulate what will happen if you and your group buy 100 boxes of these cookies. How many times did you have to pay $3? $4?

  c. How much did you and your group pay in all for the 100 boxes of cookies from the simulation? How much would you have paid for the 100 boxes if you had paid $3.23 per box? Does Mr. Rossides' proposal seem fair? Explain your answer.

their results to demonstrate the range of possible answers.

b. If students paid $0.75 for each drink, then $60 \times \$0.75$ would be $45 for the 60 drinks. In the long run, this seems to balance out.

4. **a.** A person would pay $4 for a random selection of 1 to 23. This person would pay $3 for a random selection of 24 to 100. The rule can be summarized by the following diagram in which $N = 23$ for this problem:

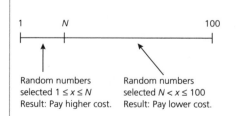

b. Direct students to obtain 100 random numbers within the range of 1 to 100. If students are directed to use the TI-83, then the following option would generate the 100 numbers: **randInt(1,100,100)**

STUDENT PAGE 24

The number of times a student would pay $4 is the number of random numbers between 1 and 23. The number of times a student would pay $3 is the number of random numbers between 24 and 100.

c. Answers will vary for the total amount paid by each student. You can expect 23% of the numbers to be between 1 and 23, and 77% of the numbers to be greater than 23. If this is the exact spread from the simulation, then the total cost from the simulation is 23($4) + 77($3) = $92 + $231 = $323.

If students paid $3.23 per box, then the total cost for 100 boxes is also $323. Comparing results from the simulation among the students will indicate that some students paid more and some paid less. The overall results of the process, however, appear fair.

5. a. The probability of guessing the correct answer would be 0.50, or 50%, as there are only two options to a correct answer, true or false.

b. Direct students to obtain a sample of 5 random values of 0 or 1. If students use the TI-83, then the 5 values can be obtained by the following expression: **randInt(0,1,5)**

If students are using a random-number table, then the selection of a 0 or 1 needs some clarification. One of several options would be to let a 0 result from an even digit and a 1 result from an odd digit.

Whatever method is used, students should be directed to determine the total number of correct and incorrect responses from their 5 random selections.

c. Students repeat the process from part b. If a calculator is used, the process can be repeated by simply selecting **ENTER** after the 5 selections are displayed. This will repeat the function of selecting another set of 5 numbers.

Many Outcomes per Trial

You decide to guess at all of the answers on a true-false test. To investigate your chances of correctly answering more than half of the questions, use a simulation.

5. There are five questions on the test, so the simulation must be designed to simulate guessing the answers to the five questions and keep track of the number of correct answers.

a. What is the probability of guessing the correct answer for any one question? What could you use to simulate this probability?

b. One way to simulate guessing the correct answer to a true-false question is to randomly choose a 0 or a 1, with 0 representing a correct answer. Since there are five questions, you will need to select five random numbers. Use a graphing calculator or some other process to select five random numbers from 0 and 1 and count the 0s, or correct answers. Record this number of 0s.

c. Repeat the procedure for a total of 15 trials. These trials represent taking the test 15 times. Record the number of correct answers for each trial.

d. Combine the number of correct answers from all members of your group. Using a number line similar to the one below, construct a graph of your group's results by writing an X above the number of correct answers. This graph is the distribution of the number of correct answers.

Distribution of Correct Answers

```
    0    1    2    3    4    5
        Number of Correct Answers
```

Your distribution of numbers of correct answers may look something like the one shown. In this example, three or more correct answers were obtained 23 times out of 50, or 46% of the time. The average number of correct answers per trial is 2.36. The average, also known as the *expected value,* was found by multiplying the number of correct answers by the frequency, finding the sum of these products, and then dividing by the total number of trials.

---

d. The problem directs students to develop a type of line plot of the number of correct answers from each trial. Students can either use the 15 trials they simulated or the combined trials developed with other students to develop this histogram. Place an X over 0 for each trial in which no correct answers were obtained or all the random numbers were 1. Similarly, place an X over the 1 for each trial in which one correct answer was obtained or 1 of the five random numbers

was 0 and the rest of the values were 1. Repeat this process for 2, 3, 4, and 5.

The example developed in the remaining explanation of this problem is an *average* value of selecting correct answers from 5 questions using the 50 trials summarized. This value is an average of the number of correct answers obtained from the 50 trials, an *expected* value of the number of correct answers.

STUDENT PAGE 25

(5) e. This question directs the students to examine the more precise histogram of the previous data. The number of trials in which 3 or more correct answers were guessed is 14 + 8 + 1 or 23 trials. 23 trials out of the 50 trials conducted would indicate a probability of 46%, the same probability discussed previously.

f. The average number of correct guesses would again be 2.36.

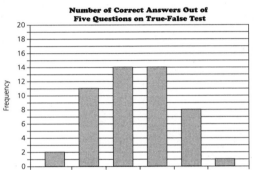

Distribution of Correct Answers

Number of Correct Answers	Frequency	Product
0	2	0
1	11	11
2	14	28
3	14	42
4	8	32
5	1	5
5	1	5
Totals	**50**	**118**

Average = $\frac{118}{50}$ = 2.36

Number of Correct Answers Out of Five Questions on True-False Test

e. Approximate the probability that you would correctly answer three or more questions by guessing.

f. What is the average number of questions correctly answered per trial? Show your work.

The average in Problem 5f is an approximation of your expected number of correct answers when you take the test by guessing.

Practice and Applications

6. a. The number of selections per trial is changed to 10.

b. Students follow the directions as indicated. One trial would be represented by the following:
randInt(0,1,10)

Repeated selections of **ENTER** would generate the number of trials requested by this question. The problem is rather tedious if not developed within small groups.

c. The graph requested can be developed by placing an X over the number of correct answers obtained per trial. A historgram could also be developed by entering the results within the LISTS of a graphing calculator. Enter the values 0–10 in L1 and the frequencies of each outcome in L2. Review the steps outlined in Lesson 1 to produce a histogram. The general shape of this histogram should build up to the value 5. Advanced studies of this simulation would develop a recognition of the binomial distribution.

STUDENT PAGE 26

Summary

Many probability questions ask how many times a specific outcome—that is, a success—will occur in a fixed number of opportunities. An example of this situation includes how many questions a student got correct on an exam, how many hits a baseball batter got in his or her last 10 trips at-bats, or how many babies among the 20 born in a local hospital were girls. The distributions of the number of successful outcomes can be simulated by the following steps:

  i. Determine the probability for each random selection.

  ii. Determine how to construct an event having this probability from random numbers.

  iii. Determine how many selections are in a trial of the simulation.

  iv. Run many trials of the simulation, recording the number of successes for each.

  v. Make an appropriate graph of the results for all trials.

  vi. Use the simulated distribution of outcomes to answer questions about probabilities and averages or expected values.

Practice and Applications

6. Suppose you are now guessing your way through a ten-question true-false test. Conduct a simulation for approximating the distribution of the number of correct answers.

  a. Does the number of random selections per trial change? If yes, what is it now?

  b. With your group, conduct at least 50 trials and record the number of correct answers for each trial. Remember, one trial means you have simulated taking the ten-question test one time.

  c. Construct a graph of the results using a number line similar to the one below.

Distribution of Correct Answers

```
  0   1   2   3   4   5   6   7   8   9   10
              Number of Correct Answers
```

STUDENT PAGE 27

(6) d. The probability of obtaining more than half (or 6 or more correct answers) should be close to the answer obtained from the previous simulation of obtaining 3 or more correct answers. Taking a long or a short test should not make much of a difference in obtaining more than half of the answers correct. However, a short test is much more likely to result in a passing score (say 75% or higher) or a very poor score (25% or lower) than is a long test, which will probably have a score very close to 50%.

e. Students should obtain this expected number by following the same steps outlined in the previous examples. If a TI-83 graphing calculator is used, the students could enter the values 0–10 in L1 and the frequencies of each number of the 50 trials in L2. Set L3 equal to L1 *L2. The sum of the values in L3 divided by 50 represents the expected number of correct answers for the 50 trials. The sum of any list of a TI-83 can be obtained by selecting **2nd LIST** and the **MATH** option from that window. Option **5,** or **5:sum(,** can be used to determine the sum of the specified list, or **sum(L3)**.

7. a. The probability of guessing a correct response would be 1 out of 5 = 0.20 = 20%.

b. For a 10-question test, you would need to select 10 random selections of 1, 2, 3, 4, or 5.

c. Follow the directions as indicated in this problem. Again, this is a rather tedious problem that is best developed in small groups.

d. As with the previous problem, this graph can be developed by hand or by using the **LIST** options of a graphing calculator.

e. The sum of the correct answers representing 6, 7, 8, 9, and 10 will be a very small number. The proba-

---

d. Approximate the probability of getting more than half the questions correct. How does this compare with the answer to Problem 5e? If you had to guess at the answers on a true-false test, would you want to take a long test or a short one? Explain your answer.

e. What is your average or expected number of correct answers when guessing on this test? Explain how you found your answer.

7. Suppose you are now guessing your way through a ten-question multiple-choice test in which each question has five possible choices, only one of which is correct. Conduct a simulation for approximating the distribution of the number of correct answers.

a. You must choose a random number to correspond to guessing on this multiple-choice test. What is the probability of guessing the correct answer to any one question?

Since there are five choices, you could randomly select a 1, 2, 3, 4, or 5, with a 1 representing the correct answer.

b. For each trial, how many numbers from 1 to 5 must you randomly select?

c. With your group, conduct at least 50 trials and record the number of correct answers for each trial.

d. Make a graph of the results, using a number line similar to the one below.

Distribution of Correct Answers

```
0   1   2   3   4   5   6   7   8   9   10
```
Number of Correct Answers

e. What is the approximate probability of getting more than half of the answers correct?

f. What is the average number of correct answers per trial? Show your work.

g. How many answers would you expect to get correct when guessing your way through this test?

---

bility of getting more than 5 answers correct is dramatically different in this problem. Although students are not provided the entire explanation of a binomial distribution, this result should be explained by highlighting the increased number of possible guesses that are not correct. The explanation provided in the conclusion of this lesson indicates this probability is approximately 2%.

f. The average number of correct answers per trial will be developed

in the same way as in the previous examples. If the **LIST** options of a graphing calculator are used, L1 contains the values 0 to 10 and L2 contains the frequencies of each response. L3 should be designed to contain the product of L1 and L2, or L3 = L1 * L2. The average of correct responses would then be the sum of L3 divided by the number of trials simulated (50). This result should be close to 2 correct answers per trial.

g. Essentially this question is rephrasing the meaning of the average calculated in part f. Students would expect to get approximately 2 of the 10 questions correct if they guessed the answer for each question. (Note: This value is also equal to 10 times the probability of getting one question correct, or 0.20.)

Extension

8. Completion of the table is dependent on the values students developed in the previous examples. The summary of the lesson, however, is that the mean of the correct answers is approximately equal to *n* times *p,* or *m = np.*

This summary is again developed in later studies of the binomial distribution. A more advanced treatment of probability uses this observation extensively. (See the *Probability Models* module.)

STUDENT PAGE 28

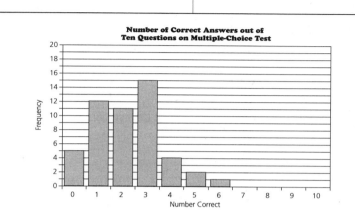

Number of Correct Answers out of Ten Questions on Multiple-Choice Test

Your simulation of the multiple-choice test may look something like the one below. Here the chance of getting six or more answers correct is only $\frac{1}{50}$, or 0.02. The average number of correct answers per trial is 2.2.

Extension

8. For each of the simulations in this lesson, identify the following variables:

n = the number of selections per trial

p = the probability of obtaining a specific outcome—that is, a success—on any one selection

m = the average, or **mean,** number of successes per trial in the simulation

	n	*p*	*m*
Five-Question True-False Test	——	——	——
Ten-Question True-False Test	——	——	——
Ten-Question Multiple-Choice Test	——	——	——

Do you see any relationship between *m*, calculated from the simulation data, and *n* and *p*? Write a formula for *m*, using the variables *n* and *p*.

LESSON 4

Waiting for Success

Materials: random-number table (optional), Lessons 3 and 4 Quiz (optional)
Technology: graphing calculator
Pacing: 1 class period

Overview

In this lesson, students continue to conduct simulations. The simulations in this lesson are different from those in Lesson 3, in that the number of trials is not fixed. The simulations in this lesson ask the question "How long will I have to wait until the first success occurs?" In the first investigation, the probability of a success is 50%, while the second investigation illustrates an example in which the probability of a success is 10%. As in Lesson 3, the steps of the simulation are presented in the summary of the lesson. It is important that students be able to follow these steps and write a clear explanation of how they conducted their simulations, what data were gathered, and what conclusions are reached.

Teaching Notes

The first investigation could be simulated in a number of different ways. Students could toss coins, designating a head as finding a customer with two or more accounts. Students could also use a random-number table or a graphing calculator. As in the first three lessons, it is again important that students collect data from the class and display this data in a graph to see the centering and the overall shape of the distribution.

Follow-Up

This lesson can be extended by changing the probability of a success and changing the number of success that need to occur. For example: If a basketball player makes 60% of the free throws attempted, on the average how many free throws would this player have to shoot in order to have 4 successes?

Solution Key

Discussion and Practice

1. Answers will vary. Some students will likely indicate that one selection is needed; others may comment that at least two selections are needed. This problem is provided to introduce the topics of this lesson.

STUDENT PAGE 29

LESSON 4

Waiting for Success

How many children will a couple have before the first girl arrives?

How many times will I have to play a game with a friend before I finally win? How often will I have to take my driver's test before I pass?

How many medications must a physician try with a sick patient until she finds one that works?

How many times have you considered questions like these? Waiting-time problems such as those above are common in probability and statistics. The problems associated with these questions all have common characteristics that will be invested in this lesson.

INVESTIGATE

Waiting for a Customer

The president of a local bank knows that 50% of the bank's account customers each have more than two accounts with the bank. The president would like to discuss the bank's services with one customer who has more than two accounts.

Discussion and Practice

1. If the president randomly selected customers, how many customers would he or she have to contact before finding one that has more than two accounts?

OBJECTIVES

Recognize probability problems that are of the form "How long will I have to wait until the first success occurs?"

Simulate the distribution of the number of trials until a success is achieved.

Use simulated distributions to make decisions.

STUDENT PAGE 30

2. The simulation discussed in the problem is an interesting variation of the simulations developed in Lesson 3. Stress that the number of selections made to obtain a success is a trial. As a result, for one trial several tosses of the coin or several random selections are needed before the success is achieved. In another trial, the success might be reached with the first selection.

If a graphing calculator is used, then 0 could be used to indicate the selection of a customer with 2 or more accounts. One method is to set randInt to **randInt(0,1).**

Hit **ENTER.** If the output is 0, then record this trial as 1. If the outcome is 1, then hit **ENTER** again. Continue to hit **ENTER** until a selection of 0 is obtained. Record the number of times it took until the 0 was selected. This number represents one trial of the simulation. Repeat the process for 30 trials.

a. As in Lesson 3, students should record the results as directed in the problem. Again, a histogram of this data can be displayed using a graphing calculator and the **LIST** options. The number of customers contacted should be in L1. Be certain that the maximum number used in this list represents the maximum number obtained in the simulation. The frequency of each number in L1 should be recorded in L2. Review the Lesson 1 steps for developing a histogram of L2 and L1.

b. Answers will vary. It is anticipated that this number will not be too great; however, various simulations might produce an unusual set of outcomes.

c. Answers will vary depending on the data set developed by the students.

A simulation can be designed to help find the average number of customers that the president would have to contact. Since 50% of the customers have more than two accounts each, you could flip a coin and designate a head as a customer with more than two accounts and a tail as a customer with one or two accounts.

In Lesson 3, you knew the number of trials; in this problem, however, the number of trials is unknown. It is, in fact, the answer to the question "How many randomly selected customers would the bank president have to contact before he or she finds a customer with more than two accounts?"

2. Conduct 30 trials of the simulation. One trial is flipping a coin until you observe the first head and recording the number of times you flipped the coin, representing the number of customers who had to be contacted.

a. Plot the results of the 30 trials.

Waiting for a Customer

```
0  1  2  3  4  5  6  7  8  9  10  11  12  13  14  15  16
```
Number of Customers Contacted

b. Use the results of your simulation to determine the greatest number of customers that the bank president had to contact.

c. Find the average number of customers who had to be contacted. Show how you found this average.

Waiting for Blood

The manager of a blood bank needs a donor with blood type B+ (B positive) and knows that about 10% of the donors have type B+ blood. Simulating the distribution of the waiting time until the first B+ donor shows up can help the manager see how long she might have to wait. The simulation can help her decide whether or not to issue a special call for more donors.

3. Set up the simulation for this waiting-time problem.

a. Assume you will be selecting a random number from 1 to 100. Which numbers will represent a donor with B+ blood?

---

3. **a.** Answers will vary; however, the most common answer is anticipated to be the numbers 1 to 10, as this set represents 10% of the total possible numbers.

STUDENT PAGE 31

b. Select a random number from the numbers 1 to 100. Count the number of selections until a number from 1 to 10 is selected. This process represents one trial. If the numbers selected are obtained by using a graphing calculator, generate each number using the following settings of randInt:

randInt(1,100)

c. Students are directed to follow the directions described in the problem.

d. Students should be directed to plot the data as indicated. As mentioned in previous problems of this type, students could also develop this graph with a graphing calculator.

e. The shape of this graph peaks to a high value and then decreases. Consider other descriptions developed by students supporting the shape of their particular data set.

f. To determine the average number of donors, students need to multiply each observed number of donors by the frequency of that number. The sum of these values for each observation is the total number of observed donors. This total is divided by the number of trials, 30. This average will represent the average number of donors selected per trial. If **LISTS** were used to store this data, then the average could again be determined by finding the sum of the list representing the number of donors divided by the 30 trials.

g. Determine the total number of trials it took to obtain 15 or more donors. Divide this number by the total number of trials, 30.

b. Describe how to conduct one trial.

c. Conduct at least 30 trials of this simulation. Record the number of donors up to and including the first B+ donor for each simulated trial.

d. Plot the outcomes of the 30 trials.

Waiting for Blood

0 1 2 3 4 5 6 7 8 9 10 11 12 13 14 15 16 17 18 19 20 . . .
Number of Donors Before First B+ Donor

e. Describe the shape of this distribution.

f. On the average, how many donors must show up before the first donor with B+ blood shows up?

g. What is the approximate probability that it will require 15 or more donors to find the first donor with B+ blood?

Summary

Many probability problems are of the form "How many times must this event be repeated before success is achieved?" The distribution of the number of repetitions until the first success can be simulated by following these steps:

i. Determine the probability for the basic random event.

ii. Determine how to construct an event with this probability from random numbers or some other randomization device.

iii. Determine how many repetitions make up one trial; this is a key step.

iv. Run many trials of the simulation, recording the number of repetitions in each.

v. Make an appropriate plot of the results for all trials.

vi. Use the simulated distribution of outcomes to answer questions about probabilities and expected values.

Practice and Applications

4. a. You would need to generate the random numbers 1 to 6. Using the graphing calculator, this could be generated by **randInt(1,6).**

A trial would consider how many selections of random numbers are needed to obtain a 6. Consider the following selections of random numbers:

i. 3 3 4 4 4 5 4 5 6

It took 9 selections of random numbers from 1 to 6 until a 6 was obtained. Therefore, for this trial, the outcome was 9.

ii. 1 4 3 4 5 1 6

It took 7 selections of random numbers from 1 to 6 until a 6 was obtained. Therefore, for this trial, the outcome was 7.

b. Students should develop this problem as directed.

c. Results will vary depending on the random numbers obtained by students. Generally, the shape of the graph will quickly peak and then show a gradual decline. The following graph is a representative set of 30 trials using random numbers generated with a TI-83.

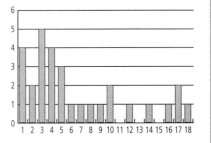

d. For the data in part c, the table on page 39 represents the steps to determine the average number of rolls to begin a play.

The average number of tosses is

sum $\frac{L3}{30} = \frac{198}{30} = 6.6$.

STUDENT PAGE 32

Practice and Applications

Many board games that involve the rolling of a number cube provide an advantage to the player who can roll a 6 fairly often. Suppose you are playing a game that requires rolling a 6 before you can take your first turn.

4. Design a simulation to find the number of rolls you might expect to make before you can take your first turn.

   a. Describe what random numbers you used, and describe how to conduct one trial.

   b. Record the results for the 30 trials, and construct a plot of the results.

   c. Describe the shape of this distribution.

   d. Over many plays of the game, what is the average number of rolls needed to begin play?

   e. What is the approximate probability that you will need to roll the die more than 10 times before you can begin play?

5. According to the United States Bureau of the Census, about 12% of American families have 3 or more children. The Gallup Organization wants to find a randomly selected family with 3 or more children in order to test a questionnaire on family entertainment.

   a. Design a simulation that will help the Gallup Organization determine the number of families needed in order to find one with 3 or more children. Write a detailed description of how you conducted your simulation. Your description should include a graph and any calculations.

   b. Do you think that the Gallup Organization will need to select more than 20 families to find one with 3 or more children? Explain your answer.

Extension

6. Problem 3 asks about the waiting time until the first donor with B+ blood arrives at the blood bank. Suppose the blood bank needs two B+ donors.

   a. Design a simulation that will help determine the number of donors that must arrive before two B+ donors arrive at the blood bank. Write a detailed description of how

---

e. Answers will vary according to the values simulated by the students. The probability of rolling more than 10 times using the above example is found by calculating the sum of the number of times, or frequency, of tossing 11 to 18 times divided by 30 trials;

that is, $\dfrac{0 + 1 + 0 + 1 + 0 + 1 + 2 + 1}{30}$

$= \dfrac{6}{30} = 0.2$, or 20%.

Number of Tosses of Die	Frequency	Product L3= L1*L2
L1	L2	L3
1	4	4
2	2	4
3	5	15
4	4	16
5	3	15
6	1	6
7	1	7
8	1	8
9	1	9
10	2	20
11	0	0
12	1	12
13	0	0
14	1	14
15	0	0
16	1	16
17	2	34
18	1	18

5. **a.** The selection of random numbers could be based on the following:

Generate random numbers from 1 to 100. Numbers 1 to 12 represent a family of 3 or more children. Numbers 13 to 100 represent families with 0 to 2 children. As with the other simulations of this section, the students need to record the sections until a family of 3 or more—a number from 1 to 12—is selected. Students should be directed to design their experiment so that at least 30 trials are conducted. This should allow them, in Problem 7, to compare the results with the previous examples.

The probability theories behind each simulation are more involved than can be adequately summarized at this point. If a generally accurate summary of the simulation is produced, then the average number of trials to pick a success multiplied by the probability of a success should be close to 1. Most of the simulations will indicate this

relationship. For this experiment, therefore, the average number of selections needed to obtain a family of 3 or more would be approximately 8 or 9 selections.

b. Given the average number of selections is approximately 8 or 9 selections, the selection of 20 families is a high estimate.

Extension

6. **a.** Select random numbers from 1 to 100. Let the selection of a number from 1 to 10 represent a person with type B+. In this simulation, a trial will be defined as the number of selections until two numbers within the range of 1 to 10 are picked. When the second random number within this range is picked, the number of selections made represents the value for that trial. This second selection makes the simulation more involved and rather tedious. Students should work in small groups.

STUDENT PAGE 33

(6) **b.** Guide students to carry out this simulation for at least 30 trials. Use the results of the 30 trials to determine the average number of trials before 2 donors are selected.

c. Students should use the results of simulations for this problem.

7. As indicated in Problem 5, the general relationship of the mean number of picks and the probability of a success is $mp = 1$.

The simulations designed in this lesson should generally support this observation.

you conducted your simulation. Your description should include a graph and any calculations.

b. How many donors should the blood bank have to check before finding the two B+ donors?

c. How does this waiting-time distribution compare to the one in Problem 3?

7. For each simulation in this lesson, identify the following variables:

p = the probability of obtaining the outcome of interest on any one selection

m = the mean number of repetitions up to and including the first success

Do you see any relationship between m, calculated from the simulation data, and p? That is, can you write the expected value of the number of repetitions to achieve success as a function of p?

Assessment for Units I and II

Materials: *Activity Sheets 9* and *10*
Technology: graphing calculator
Pacing: 1 class period or homework

Overview

This assessment enables you to evaluate students' understanding of Lessons 1–4. Problem 1 covers material from Lessons 1 and 2. Problems 2–4 cover material from Lessons 3 and 4.

Teaching Notes

As your students work through the problems, stress the need for them to be clear in showing their work and their method of solving the problem.

Follow-Up

After students have completed the assessment, collect class data on Problem 3. Displaying the distribution of the number of A+ blood types will give students another opportunity to summarize a simulation based on a large number of trials. If students know their own blood type, you might construct a distribution of the students' blood types and compare it to the simulated distribution.

STUDENT PAGE 34

Solution Key

1. **a.** See table on page 41.

Assessment for Units I and II

OBJECTIVE
Apply the concepts of simulation to answer probability questions.

1. A student was interested in the percent of M&Ms® produced that are blue. The student conducted an experiment and collected the following data.

M&M Number	Outcome (Color)	M&M Number	Outcome (Color)
1	Red	9	Brown
2	Brown	10	Green
3	Red	11	Yellow
4	Blue	12	Blue
5	Green	13	Brown
6	Brown	14	Brown
7	Blue	15	Orange
8	Red		

a. Complete the following table, or use the table on *Activity Sheet 9*.

M&M Number	Outcome	Cumulative Number of Blue M&Ms	Relative Frequency of Blue M&Ms
1	————	————	————
2	————	————	————
3	————	————	————
.	————	————	————
.	————	————	————
.	————	————	————
15	————	————	————

b. Use the data from the 15 M&Ms to construct a line graph of the relative frequency of blue M&Ms. Use a grid like the following or the grid on *Activity Sheet 10*.

M&Ms is a registered trademark of M&M/Mars, a division of Mars, Inc.

M&M Number	Outcome	Cumulative Number of Blue M&Ms	Relative Frequency of Blue M&Ms
1	Red	0	$\frac{0}{1} = 0.00$
2	Brown	0	$\frac{0}{2} = 0.00$
3	Red	0	$\frac{0}{3} = 0.00$
4	Blue	1	$\frac{1}{4} = 0.25$
5	Green	1	$\frac{1}{5} = 0.20$
6	Brown	1	$\frac{1}{6} \approx 0.17$
7	Blue	2	$\frac{2}{7} \approx 0.29$
8	Red	2	$\frac{2}{8} = 0.25$
9	Brown	2	$\frac{2}{9} \approx 0.22$
10	Green	2	$\frac{2}{10} = 0.20$
11	Yellow	2	$\frac{2}{11} \approx 0.18$
12	Blue	3	$\frac{3}{12} = 0.25$
13	Brown	3	$\frac{3}{13} \approx 0.23$
14	Brown	3	$\frac{3}{14} \approx 0.21$
15	Orange	3	$\frac{3}{15} = 0.20$

(1) b.

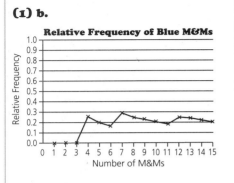

c. Based the line graph, the percent of blue M&Ms converges to a position close to 0.20, or 20%.

d. Consider the following as one possible extension of the graph in part b.

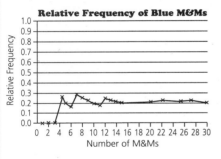

Primarily note the convergence of a student's estimate of the percent of blue M&Ms. The differences in the values should not be as pronounced as the first 15 selections.

e. This answer should be based on a student's estimate of the percent of blue M&Ms. If this estimate is 20%, then for a bag of 75 M&Ms, a student would predict 0.20 × 75 M&Ms = 15 blue M&Ms.

2. **a.** Consider the selection of random numbers within the range of 1 to 100. Let the selection of a number from 1 to 80 represent a successful free throw and a selection of a number from 81 to 100 as an unsuccessful free throw. Let 10 selections represent a typical game. If a TI-83 is used, the following setting would simulate a game:

STUDENT PAGE 35

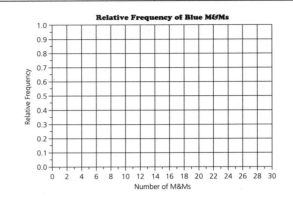

c. What is your estimate for the probability of randomly selecting a blue M&M?

d. If 15 more M&Ms were sampled, how would this change the line graph? Extend your line graph to show your answer.

e. Predict the number of blue M&Ms in a bag containing 75 M&Ms.

2. The star free-throw shooter on the girls' basketball team makes 80% of her free throws. She takes about 10 such shots per game.

a. Design and conduct a simulation that shows the approximate distribution of the number of successful free throws per game for this player. Show all the steps of the simulation, including what random numbers you used, what constituted one trial, and a plot of the resulting data.

b. What is the approximate probability that she will make more than 80% of her free throws in any one game?

c. How many free throws should she expect to make in a typical game?

d. Over the course of a 15-game season, how many free-throw points should this player expect to score?

e. Are there any assumptions built into the simulation that might not be realistic? Explain your answer.

randInt(1,100,10)
The following random numbers are changed to successful (Yes) or not successful (No) free throws made in 1 game.

7   69   66   17   88   10   90   24   51   62
↓   ↓   ↓   ↓   ↓   ↓   ↓   ↓   ↓   ↓
Yes Yes Yes Yes No   Yes No   Yes Yes Yes

The above outcome is recorded as 8 free throws and represents 1 trial.

Students should simulate several games. If 30 games (or trials) were simulated, consider the following representative results:

Game Number	Number of Free-Throws Made (frequency)
0	0
1	0
2	0
3	0
4	0
5	0
6	1
7	4
8	8
9	10
10	7

A plot of these data follows.

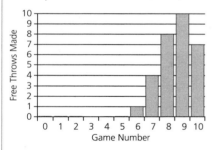

b. Based on the representation of the data in part a, an approximate probability that she will make more than 80% of her free throws per game would be the probability of making 9 or 10 free throws, or $\frac{10 + 7}{30} = \frac{17}{30}$ or $\approx 57\%$.

Answers will vary according to the results simulated by the student.

c. The expected number of free throws made per game can be found by dividing the total number of free throws made in the 30 games by 30. For the data represented above, this result is

$$\frac{(6 \times 1) + (7 \times 4) + (8 \times 8) + (9 \times 10) + (10 \times 7)}{30}$$

$$= \frac{258}{30} \approx 8.6.$$

Therefore, this player averaged 8.6 successful free throws per game.

Answers will vary according to the specific simulations developed by the students. Note that the above average is close to the value np or (10 games)(0.8) = 8 successes per trial. This represents a mean or average based on 10 free throws in a trial and an 80% probability of success.

d. Over the course of 15 games, you would expect this player to make 15×8 free throws per game = 120 free throws.

As each free throw represents 1 point, a student might estimate the player will make 120 points from free throws during the season.

e. Answers will vary; however, it is apparent that 10 free throws per game might not be realistic.

STUDENT PAGE 36

3. **a.** A simulation might be developed in which random numbers from 1 to 100 are again generated. A trial would be represented by 20 selected numbers. Numbers in the range of 1 to 33 would represent A+ blood, while numbers in the range of 34 to 100 would represent all other types of blood. This simulation should be developed with several trials. A chart and a graph should be used to evaluate a student's steps in developing the simulation. Given the summaries developed in the lessons, the average expected number of donors with A+ blood per day would be *np* or 20 × 0.33 = 6.6 patients with type A+ blood per day.

b. The probability of getting 10 or more from this simulation is small. Use the probability from a student's simulation to evaluate this answer. From the simulation, the student would determine the number of times that 10 or more donors per day were found and divide this by the number of trials conducted in the simulation. It is not likely that 10 or more donors will be found during one day.

c. This result should be close to the average expected number of donors, or 6.6 donors per day.

4. This problem should be modeled after the problems in Lesson 4. Students can again let random numbers from 1 to 100 represent patients. As in Problem 3, a selection of a number from 1 to 33 would represent an A+ donor, and a selection of 34 to 100 would represent a donor of the other types of blood. In this problem, however, a trial would be defined as the number of selections until a number within the range of 1 to 33 is selected. The simulation should

3. About 33% of the people who come into a blood bank to donate blood have type A+ blood. The blood bank gets about 20 donors per day.

   a. Set up and conduct a simulation that shows the approximate distribution of the number of A+ donors coming into the blood bank per day. Be sure to show all your work for each step of the simulation.

   b. If the blood bank needs ten A+ donors tomorrow, is the bank likely to get them? Explain your answer.

   c. How many A+ donors can the blood bank expect to see each day?

4. Refer to the A+ blood donor in Problem 3. How would the setup of the simulation change if the blood bank were interested in the number of donors who must be seen until the first A+ donor shows up? Outline the simulation; however, you need not carry it out.

involve repeating the trials a significant number of times to evaluate the questions posed in the problem.

Data Tables and Probability

LESSON 5

Probability and Survey Results

Materials: list of students in school, *Activity Sheets 11* and *12,* random-number table (optional)
Technology: graphing calculator (optional)
Pacing: 2 class periods with extra time for conducting the survey

Overview

In the first four lessons, we have established that with enough trials the experimental probability approaches the theoretical probability of an event. This lesson asks students to apply this idea to analyzing survey results. As the sample size increases, the percent of those answering "yes" should approach the percent of the entire population who would answer "yes."

The lesson begins by briefly discussing a random sample. The students are asked to take a random sample of students in school. The results are tallied in a chart similar to those completed in Lessons 1 and 2. A line graph of the results is constructed and conclusions are drawn.

The rest of the lesson presents results from surveys found in newspapers and magazines. Each problem presents the survey results in a different format. Problem 5 presents the data as frequencies and students convert to percent. In Problem 6, the data are presented in a bar graph with the percent given. In Problem 7, the data are given as percents and students are asked to construct a bar graph. Problem 8 presents the data as cumulative percents.

Teaching Notes

Be sure to discuss the two questions in the Investigate paragraph. The lesson continues by asking students to take a random sample of the student body. This should be a class project with each student responsible for asking only one or two students in the school. When the results are compiled, it is important that students see that the relative frequency of yeses changes as the number of surveys counted increases. You may wish to have the students ask a different question and survey only one grade level.

The rest of the lesson presents students with different survey results. As students convert the frequency to relative frequency, it is important that they realize that the sum of the relative frequencies should be 1. Problems 8 and 10 present cumulative percents. Caution your students to pay particular attention to what the given percents represent.

Follow-Up

Problem 11 asks the students to find examples of surveys from newspapers or magazines. When your students find examples, they should list the source and date and their summaries should be well-written. The questions that they write could be used for a quiz.

STUDENT PAGE 39

Probability and Survey Results

Do you eat breakfast on a regular basis?

Do other members of your family ever skip meals?

Students at Rufus King High School in Milwaukee, Wisconsin, as part of a research project, conducted a survey asking the question "Do you eat breakfast at least 3 times a week?" Instead of contacting every student in the school, the students took a *random sample* of all the students.

After the survey was conducted, the students used the survey results to draw conclusions about the entire student body. As they organized and analyzed the data from the survey, they noticed similarities between their survey results and the experiments conducted in Lessons 1 and 2. In this lesson you will investigate these similarities and use results from a survey to draw conclusions about the probability of an event.

OBJECTIVE

Find an estimate of the probability of an event given the results of a survey.

INVESTIGATE

Breakfast Survey

A *simple random sample* is a sample chosen in such a way that every possible sample of a given size has an equal chance of being selected.

To select 50 students from the student body, the students could put the name of every student on a card, put the cards in a box, mix them up, and draw 50 cards. Since every possible sample of 50 students has the same chance of being chosen, this would be one way to take a random sample of the entire student body.

STUDENT PAGE 40

Solution Key

Discussion and Practice

1. If developing a random sample of the school is too difficult, students could survey the class. As the students analyze the data from the class, they should keep in mind some of the reasons why their class might not be a good representation of the school. The results recorded by the students in the chart should be modeled after the earlier lessons of this module.

Can you think of some other methods that the students might have used to take a random sample of the entire student body? Do you think your class is a random sample of the entire student body in your school?

Discussion and Practice

In this survey, your class will take a random sample of 30 students in your school. To ensure that you are taking a random sample, your class might give each student in school a number and then select 30 numbers at random.

1. Ask each student in the survey this question: "Do you eat breakfast at least 3 times a week?" Collect the results of the survey in a table similar to the following, also available on *Activity Sheet 11*.

Do You Eat Breakfast?

Student	Outcome (Yes or No)	Cumulative Number of Yeses	Relative Frequency of Yeses
1	————	————	————
2	————	————	————
3	————	————	————
.	————	————	————
.	————	————	————
.	————	————	————
30	————	————	————

2. Use the results of your survey to construct a line graph of the relative frequency of yeses. Use a grid like the one on page 41, also available on *Activity Sheet 12*.

STUDENT PAGE 41

2. **a.** The line graph will probably indicate some erratic movement in the beginning of the sampling. As more students' responses are recorded, the line should appear to level off, or converge, at a specific value.

b. This value will be used to estimate the probability of selecting a student from the school who indicated "yes" to the question involving breakfast.

c. The line graph provides us with a method to estimate the probability of selecting a student who answered "yes" to the breakfast question.

3. **a.** The estimate used in this problem should be the same value observed in Problem 2b.

b. A student should use his or her estimate of the probability of selecting one student who answered "yes" to determine the number out of 100 randomly selected students. If a student's estimate of randomly selecting a student who answered "yes" to this question is 25%, then the number of students in a sample of 100 who answered "yes" is $(0.25) \times 100 = 25$ students.

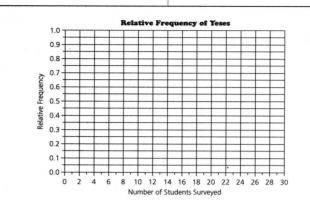

Relative Frequency of Yeses

a. As you tallied more and more surveys, describe what happened to the line graph.

b. As the number of students surveyed increased, what relative frequency does the line graph approach? Draw a horizontal line across your graph at this point.

c. What does the line graph of relative frequency tell you about the percent of students at your school who eat breakfast at least 3 times a week?

3. Use the results from the survey in Problem 2.

a. Suppose you randomly selected one student. What do you think the probability is that the student will answer "yes" to the question "Do you eat breakfast at least 3 times a week?"

b. If you randomly sampled 100 students, how many students would you expect to say that they eat breakfast at least 3 times a week?

STUDENT PAGE 42

4. **a.** The data indicate a leveling off at around 0.75, or 75%, of the responses.

b. An estimate of the probability of selecting a student who eats breakfast at least 3 times per week is 0.75, or 75%.

c. Essentially use the estimate described in part b to determine the number of students out of 850 that would be expected to eat breakfast. With an estimate of 75%, 850 students × 0.75 = 637.5; that is, approximately 638 students would eat breakfast.

4. The line graph below shows the results of the survey conducted by the Rufus King students. The graph shows the variation in the relative frequency of "yes" responses to the question "Do you eat breakfast at least 3 times a week?" Compare the results of your class survey to those of the King students.

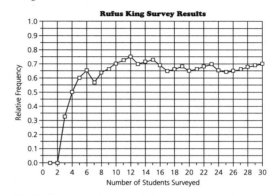

Rufus King Survey Results

The Rufus King students sampled more than 30 students. The following table summarizes the survey results after the students tallied the remaining surveys collected at the school.

Number of Students	Number of Yeses	Relative Frequency of Yeses
40	30	0.75
80	59	0.74
100	74	0.74
150	112	0.75

a. What do the data tell you about the percent of Rufus King students who eat breakfast at least 3 times a week?

b. If one King student were randomly selected, what do you think the probability is that the student will say "yes" to the question "Do you eat breakfast at least 3 times a week?"

c. If 850 students attend Rufus King High School, how many students would you expect to eat breakfast at least 3 times a week? Explain how you arrived at your answer.

STUDENT PAGE 43

5. a. See table below.

b. Based on the table, an estimate of 0.32, or 32%, is probably appropriate.

c. An estimate of 15% is appropriate.

Different Survey Results and Probability

5. A survey with a large sample size can be used to answer questions concerning the entire population under study. A poll asked a random sample of 2500 high-school students the question "When you go out on a date, who pays for the date?" The results are shown below.

Responses	Frequency
Boy	1200
Split Costs	800
Girl	75
Girls' Parents	25
Boys' Parents	25
Don't Date	375
Total	2500

The results of this survey can be used to draw some conclusions about the population of high-school students.

a. Make a table of the data with a column added for relative frequency of responses. Calculate the relative frequency for each response as a fraction, a decimal, and a percent. Find the sum of the relative frequencies.

b. Suppose that only one high-school student were randomly selected. What is an estimate of the probability that the student would say the costs are split when they date?

c. What is an estimate of the probability that a randomly selected student would say that he or she does not date?

In Problem 5, the results of the survey were given in terms of the number of people in each category. Many times, survey results are reported using the percent of outcomes in each category. These percents can also be thought of as relative frequencies. Since the survey had a very large sample, these percents can be used as an estimate for the probability of individual outcomes.

Responses	Frequency	Relative Frequency
Boy	1200	$\frac{1200}{2500} = 0.48 = 48\%$
Split Costs	800	$\frac{800}{2500} = 0.32 = 32\%$
Girl	75	$\frac{75}{2500} = 0.03 = 3\%$
Girls' Parents	25	$\frac{25}{2500} = 0.01 = 1\%$
Boys' Parents	25	$\frac{25}{2500} = 0.01 = 1\%$
Don't Date	375	$\frac{375}{2500} = 0.15 = 15\%$
Totals	2500	100%

STUDENT PAGE 44

6. a. First determine the total percent of children who play 3 or more hours per day. This involves adding the percents from the categories "3," "4–5," and "6 or More" or 6% + 4% + 2% = 12%. Then calculate 12% of 750: 0.12 × 750 = 90 children.

b. The survey indicates that 43% of the children play video games less than 1 hour a day. As this sample was rather large, 43% could be used to estimate the probability of selecting a child who plays less than 1 hour of video games per day.

c. In this problem, the percent of children from the survey who play more than 1 hour per day is 15% + 6% + 4% + 2% = 27%. Therefore, 27% could be used to estimate the probability of students who play more than 1 hour per day.

d. The sum of the percents is 99%. Rounding of the answers for each individual category explains why the percents do not total 100%.

6. A study reported that nearly 90% of children aged 9–13 play video games an average of 1.4 hours a day. The graph below shows the percent of children playing a given number of hours per day.

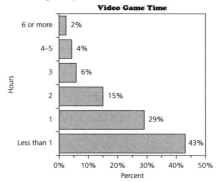

a. If the survey consisted of 750 children aged 9–13, how many of these children would say they play video games 3 or more hours a day?

b. If a child aged 9–13 is randomly selected, what is an estimate of the probability that the child plays video games less than 1 hour a day?

c. If a child aged 9–13 is randomly selected, what is an estimate of the probability that the child plays video games more than 1 hour a day?

d. What is the sum of the percents given in the graph? Why don't they add to 100%?

7. A survey reported the age distribution and the percent of people who purchased running shoes during the last year. The results shown below describe the distribution.

Age of Purchasers	Percent
Under 14 Years Old	8.3%
14 to 17 Years Old	10.5%
18 to 24 Years Old	10.0%
25 to 34 Years Old	28.5%
35 to 44 years Old	24.0%
45 to 64 Years Old	15.2%
65 Years Old and Over	3.5%

STUDENT PAGE 45

7. a.

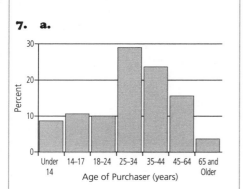

The age distributions are not equal for each of the categories represented. The first category, "Under 14," represents 14 years; the second category, "14–17," represents 4 years; and so on.

b. This question can involve a lot discussion. *Typical* is not a precise statistical term and yet is frequently used to summarize data. If *typical* refers to the most frequent buyers, then the age of the typical buyer would be 25 to 34 years old, as this group made 28.5% of the purchases. Other groupings of the data might be developed to define *typical* in other ways.

c. i. 10.5%

　　ii. 15.2% + 3.5% = 18.7%

　　iii. 10.0% + 28.5% = 38.5%

　　iv. 10.5% + 8.3% = 18.8%

　　v. 10.0% + 28.5% + 24.0% + 15.2% + 3.5% = 81.2%

8. a. The percents in this data set are based on cumulative values. As indicated in the description of this problem, each new category, for example, "less than 50 years," includes the percent of families in the previous category plus the additional percent of families.

---

a. Construct a bar graph with the ages of the purchasers as individual categories and the percent as the height or length of each bar. This graph represents the distribution of running-shoe buyers. What do you notice about the age ranges in each category?

b. Describe the age of the "typical" running-shoe buyer. Explain your answer.

c. What is an estimate of the probability that a randomly selected purchaser of running shoes is

　　i. 14 to 17 years old?

　　ii. 45 years old or older?

　　iii. 18 to 34 years old?

　　iv. 17 years and under?

　　v. at least 18 years old?

The table of the data contains all the possibilities of ages of purchases of running shoes. If you think of the percents as probabilities, then the sum of the percents gives the total of the probabilities. The sum of the probabilities of all the outcomes is exactly 100%, or 1. (Note: Due to rounding, there are times when the sum of the decimals or percents used is not exactly 1.)

8. The data below show how long cattle ranches have been in one family, as reported by the *USA Today* on July 8, 1994. The first column gives the number of years the ranch has been in one family. The second column gives the cumulative percent of ranches having been in one family fewer than the given number of years. For example, 59% of ranches have been held in one family for fewer than 50 years. This figure also includes the 21% that have been in one family for fewer than 25 years.

Number of Years Owned by One Family	Percent
Fewer than 25 years	21%
Fewer than 50 years	59%
Fewer than 100 years	89%

a. How do the percents given in this chart differ from the percents given in previous problems?

b. **i.** 59%
     ii. 100% − 89% = 11%
    iii. 59% − 21% = 38%

Practice and Applications

9. **a.** Determine the sum of the given categories: 56.9% + 21.2% + 9.0% + 9.3% + 3.5% = 99.9%. Therefore, the category "Other" is 100% − 99.9% = 0.1%.

   b. Answers will vary. Some possible suggestions include wind and solar methods.

STUDENT PAGE 46

b. If a cattle ranch is randomly selected, what is the probability that the ranch has been held in the family

 i. for fewer than 50 years?

 ii. for 100 or more years?

 iii. for 25 to 49 years?

Summary

The results of surveys can be given in many different forms. The data can be presented in terms of items as in the "Who pays on a date?" survey. To find the probabilities of these results, the data should be converted to a relative frequency given as a fraction, a decimal, or a percent.

Sometimes the results of a survey are given as a percent. The percent could represent the information about a single item as in the video-game-time data or it could be a cumulative percent as in the cattle-ranch problem. Careful attention must be given to what percents represent.

Practice and Applications

9. The table below shows the distribution of methods of generating electricity in the United States as reported by the Energy Information Administration.

Method	Percent
Coal	56.9%
Nuclear	21.2%
Gas	9.0%
Hydro Power	9.3%
Oil	3.5%
Other	

a. What is the percent for the *Other* category?

b. List some other ways by which electricity is generated in the United States.

STUDENT PAGE 47

10. Note that the percents in this problem are cumulative; that is, each new category includes the percent of the previous category.

a. The percent for 1 car is 62% − 15% = 47%.

b. The percent for 2 or fewer cars is read directly from the table, namely 91%.

c. The percent of 3 or more cars available is 100% − 91%, or 9%.

11. Analyze this problem individually. Finding information as directed should not be difficult.

---

10. According to the United States Bureau of the Census, the number of cars available to American households is given by the following cumulative percents.

Number of Cars	Cumulative Percent of Households
0	15%
1	62%
2	91%

If a household is randomly selected, what is the probability that the household will have

a. one car available?

b. two or fewer cars available?

c. three or more cars available?

11. Find a survey from a newspaper or magazine. Summarize the results of the survey and make a list of four probability questions that could be asked using the results of the survey.

LESSON 6

Compound Events

Materials: *Activity Sheet 13;* brown paper bag with numbered red, green, and blue slips of paper
Technology: calculator
Pacing: 2 class periods

Overview

This lesson begins to formulate probability rules. Students are introduced to the *AND* Rule and the *OR* Rule. In addition, two-way tables are used to analyze data.

The lesson begins by asking students to pick a movie and then decide if they liked the movie or not. This first investigation is used as a review but will be built on later in the lesson. The next investigation introduces the AND Rule. Students are encouraged to think of the word *and* as *intersection*. The data are presented in two useful forms: a Venn diagram and a two-way table. Both presentations are very important to help students conceptualize the meaning of the AND Rule.

The OR Rule is introduced using the movie data from the first investigation. Students are asked to comment on a second movie. Venn diagrams are used, along with a two-way table, to analyze the data and answer some probability questions that use the words *and* and *or*.

Teaching Notes

When collecting data for the movie survey investigation, it is important that students record the students' names. A later investigation asks the students about a second movie. The data must be collected in the same order as the data about the first movie. If you do not want to use movie data, you could ask students if they like a certain TV show, a particular song, or a

certain book. Whatever you choose, you will also need another show, song, or book for students' comments later in the lesson.

Encourage your students to use Venn diagrams to enhance their understanding of the AND Rule. This type of diagram helps students to visualize the intersection of the two events. Students are also introduced to the use of two-way tables to collect data. The use of the tables is only introduced in this lesson and will be further developed in Lessons 8 and 10.

Students will be asked to collect data similar to the hand-dominance and eye-dominance data when they encounter Lesson 9. If you collect the data now, you will wish to save it.

Follow-Up

Have students find data in a newspaper or magazine that are presented in a two-way table. Students can then write probability questions using these data. The data and questions could be shared for additional problems.

STUDENT PAGE 48

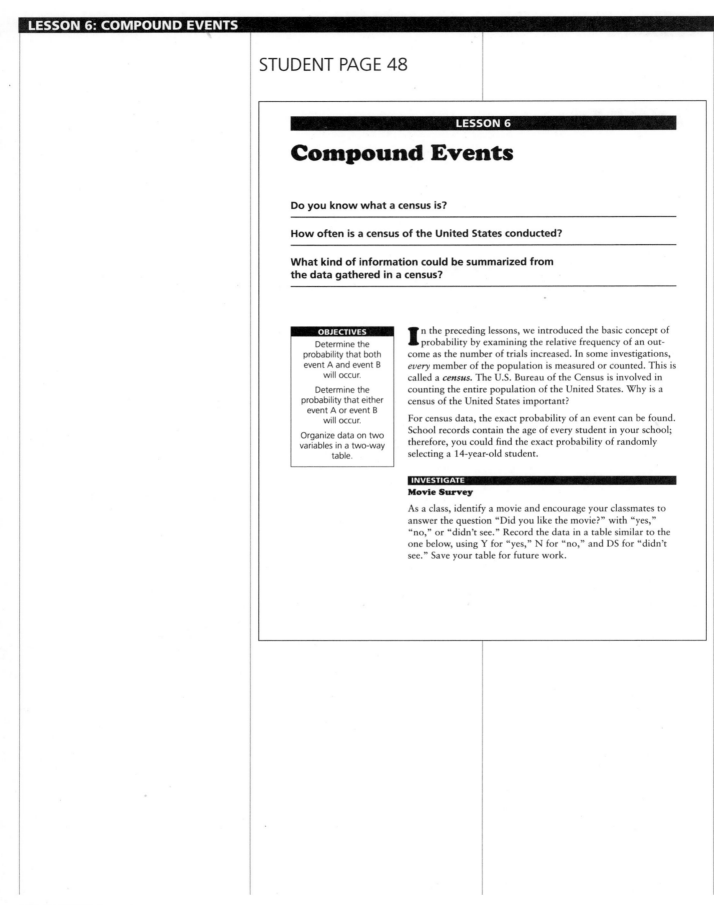

LESSON 6

Compound Events

Do you know what a census is?
_____

How often is a census of the United States conducted?
_____

What kind of information could be summarized from the data gathered in a census?
_____

OBJECTIVES

Determine the probability that both event A and event B will occur.

Determine the probability that either event A or event B will occur.

Organize data on two variables in a two-way table.

In the preceding lessons, we introduced the basic concept of probability by examining the relative frequency of an outcome as the number of trials increased. In some investigations, *every* member of the population is measured or counted. This is called a *census*. The U.S. Bureau of the Census is involved in counting the entire population of the United States. Why is a census of the United States important?

For census data, the exact probability of an event can be found. School records contain the age of every student in your school; therefore, you could find the exact probability of randomly selecting a 14-year-old student.

INVESTIGATE
Movie Survey

As a class, identify a movie and encourage your classmates to answer the question "Did you like the movie?" with "yes," "no," or "didn't see." Record the data in a table similar to the one below, using Y for "yes," N for "no," and DS for "didn't see." Save your table for future work.

Solution Key

Discussion and Practice

1. Answers will vary according to the data collected. If a more appropriate or interesting situation is suggested by students, substitute it and use the collected data.

2. a. The relative frequency of "yes" responses is the number of "yes" responses divided by the total number of responses obtained from the students in your class.

b. Similarly, the relative frequency of the "no" responses is the number of "no" responses divided by the total number of responses obtained from the students in your class.

3. a. The relative frequency found in Problem 2a represents the probability that a student will answer "yes" to the question.

b. In this case, the probability indicated in part a is exact, as the total population considered in the question is the entire group of students in your class.

c. Add the relative frequencies of the students who answered "no" and the students who answered "didn't see." The sum of the two relative frequencies would represent the probability of selecting a student who answered "no" or "didn't see."

d. The sum of the relative frequencies should be 1. If relative frequencies were converted to percents, then the sum of the percents would be 100%. Each student falls into only one of the components provided. As a result, the total number of students in the three categories would represent the entire group of students surveyed.

STUDENT PAGE 49

Student Name	Response (Y, N, DS)
_____	_____
_____	_____
_____	_____
_____	_____

Discussion and Practice

1. Combine the students' responses from your table like the one below to show the number of each response.

Movie Responses

Response	Number
Yes	_____
No	_____
Didn't See	_____
Total	_____

2. Use the results from your survey in Problem 1.

  a. Find the relative frequency of the "yes" responses.

  b. Find the relative frequency of the "no" responses.

3. Use your data from Problem 1.

  a. If a student from your class is randomly selected, what is the probability that he or she answered "yes"?

  b. Is your answer to part a an estimate of the probability or the exact probability? Explain your answer.

  c. If one student from your class is randomly selected, what is the probability that the student answered "no" or "didn't see" the movie?

  d. What is the sum of the relative frequencies for the three responses? Explain why the relative frequencies must add up to this number.

STUDENT PAGE 50

4. a. i. 5 slips of paper are even; therefore, the probability is 0.50, or 50%.

ii. 2 slips of paper are blue; therefore, the probability is 0.20, or 20%.

iii. Only 1 slip of paper is blue and even (slip blue 10); there-fore, the probability is 0.10, or 10%.

b. i. 5 slips of paper are odd; therefore, the probability is 0.50, or 50%.

ii. 3 slips of paper are green; therefore, the probability is 0.30, or 30%.

iii. Only 1 slip of paper is green and odd (green 7); there-fore, the probability is 0.10, or 10%.

c. The student with the answer $\frac{1}{10}$ is correct. Adding the probabilities involves adding selections more than once.

The *AND* Rule

4. A brown paper bag contains ten slips of paper. As shown below, five of the slips are red labeled with the numbers 1 to 5; three of the slips are green labeled with the numbers 6 to 8; and two of the slips are blue labeled with the numbers 9 and 10.

a. If one slip of paper is randomly selected, what is the probability

   i. that an even number will be drawn?

   ii. that a blue slip will be drawn?

   iii. that an even-numbered blue slip will be drawn?

b. If one slip of paper is randomly selected, what is the probability

   i. that an odd number will be drawn?

   ii. that a green slip will be drawn?

   iii. that an odd-numbered green slip will be drawn?

c. One student found the answer to the last question using the following method:

   probability of green slip = $\frac{3}{10}$

   probability of odd-numbered slip = $\frac{5}{10}$

   probability of odd-numbered green slip = $\frac{3}{10} + \frac{5}{10} = \frac{8}{10}$

   Another student disagreed and said the answer was $\frac{1}{10}$.

   Which student do you think is correct? Explain why.

Venn diagrams can be used to help find the probability that an odd-numbered green slip will be drawn. A Venn diagram is a picture showing how two or more events are related.

STUDENT PAGE 51

5. **a.** Three, as there are 3 green slips of paper

b. Five, as there are 5 slips of paper with odd numbers

c. The overlap area represents the slips of paper that are green and odd. There is only one slip of paper that fits both descriptions.

d. The area outside the circles represents the slips of papers that are not green and not odd—that is, even slips in colors other than green. There are 3 such slips of paper.

e. The following Venn diagram represents the distribution of the slips.

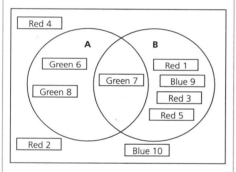

5. In the first Venn diagram below, let circle A represent the event that a randomly drawn slip is green, and let circle B represent the event that a randomly drawn slip contains an odd number.

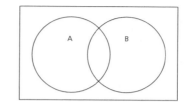

a. How many slips of paper would be represented by circle A?

b. How many slips of paper would be represented by circle B?

c. What is represented by the area where the two circles overlap? How many slips of paper are represented by this area?

d. What is represented by the area outside the two circles? How many slips of paper are represented by this area?

e. Three slips of paper are shown in the Venn diagram below and on *Activity Sheet 13*. Give the proper location for the other seven slips.

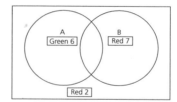

The overlap of the two circles is called the *intersection* of event A and event B. This can be represented by *A and B*. *P(A and B)* represents the probability that both event A and event B will occur.

6. a.

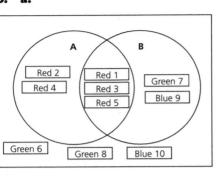

b. $P(A$ and $B)$ is 3 out of 10 = 0.3 = 30%.

7. a. 14 cars are dark green.

b. 17 cars are compact.

c. 5 cars are dark-green compact cars.

d.

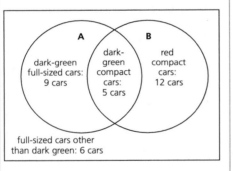

full-sized cars other than dark green: 6 cars

e. The probability of selecting a dark-green compact car is 5 cars out of 32 ≈ 0.156 = 15.6%.

STUDENT PAGE 52

6. Let event A represent the drawing of a red slip and event B represent the drawing of an odd slip.

   a. Draw a Venn diagram showing the location of each of the ten slips of paper. You can use *Activity Sheet 13*.

   b. Find $P(A$ and $B)$.

Summary

$P(A$ ***and*** $B)$ is the probability that event A and event B occur together. The word *and* indicates the ***intersection,*** or overlap, of the two events.

7. A dealership had two colors of cars in stock. When he compared the compact cars and full-sized sedans, the dealer recorded the information in the table. How many of each type of car does he have in stock?

		Color	
		Dark Green	**Red**
Type of Car	**Compact**	5	12
	Full-Sized	9	6

   a. Dark-green cars

   b. Compact cars

   c. Dark-green compact cars

   d. Draw a Venn diagram showing the relationship between dark-green cars and compact cars. Label each section of the Venn diagram and indicate how many cars are in each section. You can use *Activity Sheet 13*.

   e. If a car is selected at random, what is the probability that the car is a dark-green compact car?

The OR Rule

As a class, identify another movie and have your classmates answer the question "Did you like the movie?" Again, students may answer "yes" (Y), "no" (N), or "didn't see" (DS). Add the data to the table you made in Problem 1, as shown below.

	Movie 1	Movie 2
Student Name	**Responses (Y, N, DS)**	**Responses (Y, N, DS)**
_____	_____	_____
_____	_____	_____

STUDENT PAGE 53

8. Answers will vary according to the data collected by the class.

9. **a.** Answers will vary. The probability is the number of "yes" responses to Movie 2, divided by the total number of responses.

b. Answers will vary. The probability is the number of "no" responses to Movie 2 plus the number of "didn't see" responses to Movie 2, divided by the total number of responses.

c. Answers will vary. The probability is the number of "yes" responses to Movie 2 plus the number of "didn't see" responses to Movie 2, divided by the total number of responses.

d. The probability for this response is 100%.

10. **a.** Number of students who did not like Movie 2 and liked Movie 1

b. Number of students who did not like Movie 1 and did not see Movie 2

c. Number of students who did not see Movie 1 and liked Movie 2

d. Number of students who did not see Movie 1 and did not see Movie 2

---

8. Combine the responses for the second movie in a table similar to the one below.

Student	Response (Y, N, DS)
1	_____
2	_____
3	_____
4	_____
.	_____
.	_____
.	_____

9. If one student from your class is randomly selected, what is the probability that

a. he or she answered "yes" to the Movie 2 question?

b. he or she answered "no" or "didn't see" to the Movie 2 question?

c. he or she answered "yes" or "didn't see" to the Movie 2 question?

d. he or she answered "yes," "no," or "didn't see" to the Movie 2 question?

The table below shows a way to organize the results of the two survey questions. This *two-way table* can be very useful when investigating whether or not there is a relationship between two variables.

		Movie 2 Responses		
	Y	**N**	**DS**	**Totals**
Y	a	b	c	
N	d	e	f	
DS	g	h	i	
Totals				

Movie 1 Responses

The cell labeled "a" will contain the number of students who answered "yes" to the Movie 1 question and "yes" to the Movie 2 question.

10. Write a description of what each cell represents.

a. cell b   **b.** cell f

c. cell g   **d.** cell i

STUDENT PAGE 54

11. Answers will vary according to the data collected.

12. a. Using the labels analyzed in Problem 10, this probability is cell i divided by the total number of responses.

b. This probability is cell g divided by the total number of responses.

c. This probability is the sum of cells g, h, c, f, and i, divided by the total number of responses.

13. The second student is correct. If one adds the numbers indicated by the first student, one would be adding the students who did not see both movies twice.

14. a. Determine the total of the number of students who liked the first movie. Divide this total by the total number of students surveyed.

b. Determine the total of the number of students who did not see the second movie. Divide this total by the total number of students surveyed.

c. Use the cell labels from Problem 10 and divide the number of students represented in cell c by the total number of students surveyed.

d. Use the cell labels from Problem 10 and determine the sum of the students who liked the first movie (a + b + c). To this sum, add the additional students who did not see the second movie (f + i). Divide this second sum by the total number of students surveyed.

11. Use the class data compiled in Problems 1 and 8. Record the results of responses in a table similar to the one below.

		Movie 2 Responses			
		Y	N	DS	Totals
Movie 1 Responses	Y				
	N				
	DS				
	Totals				

12. Use your table to give the number of students in each of the following categories.

a. Didn't see either movie

b. Didn't see Movie 1 and liked Movie 2

c. Didn't see Movie 1 or didn't see Movie 2

When finding the probability that event A or event B will occur, you can use the *Addition Rule*.

$$P(A \text{ or } B) = P(A) + P(B) - P(A \text{ and } B),$$

where $P(A \text{ and } B)$ is the probability that event A and event B will occur together. You can think of the word *or* as the *union* of the two events.

13. One student found the answer to Problem 12c by adding the total number of students who said they didn't see the first movie to the total number who said they didn't see the second movie. Another student disagreed with this method because she said some students were counted twice. Do you think the second student is correct? Explain.

14. If one student from your class is randomly selected, what is the probability that the student

a. said he or she liked the first movie?

b. said he or she did not see the second movie?

c. said he or she liked the first movie and did not see the second movie?

d. said he or she liked the first movie or did not see the second movie?

e. Use the cell labels from Problem 10 and divide the number of students represented in cell f by the total number of students surveyed.

f. Use the cell labels from Problem 10 and divide (g + h + i + a + d) by the total number of students surveyed.

15. a. Answers depend on the data collected by the students. Note the wording of the sets. A represents the students who did *not* see the first movie and B represents the students who did *not* see the second movie. This description will help students understand the idea of a complement presented in the next lesson.

b.

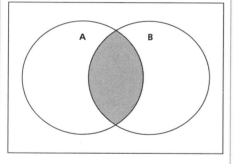

c.

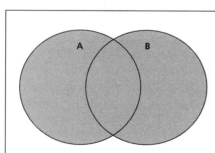

STUDENT PAGE 55

e. said he or she did not like the first movie and did not see the second movie?

f. said he or she did not see the first movie or did not like the second movie?

15. In the Venn diagram below, circle A represents students who did not see the first movie and circle B represents students who did not see the second movie.

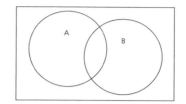

a. Copy the Venn diagram above or use *Activity Sheet 13*. Label each distinct section with the correct number of students.

b. Shade the part that represents those students who did not see either movie.

c. On another copy of the Venn diagram, shade the part that represents those students who did not see the first movie or did not see the second movie.

d. On a third copy of the Venn diagram, shade the part that represents those students who saw the first movie.

Summary

P(A or B) is the probability that event *A* or event *B* will occur. You find the probability by adding the probability that event *A* will occur to the probability that event *B* will occur and then subtract the probability that event *A* and event *B* will occur together. You can think of the word *or* as the *union* of the two events.

d. Students should shade all the regions outside circle A.

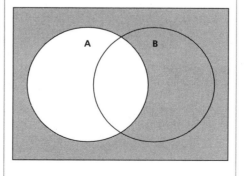

Practice and Applications

16. When examining the design of the Venn diagram, students will struggle with attempting to make C another circle. If you make the assumption that all the students are either left-hand dominant or right-hand dominant (valid based on the data), then C is not another circle, but the region outside circle A. This may challenge the students.

a.

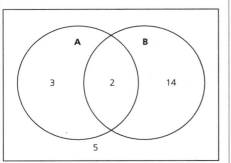

b. i. $P(C) = \dfrac{19}{24} \approx 0.791 = 79.1\%$

ii. $P(B) = \dfrac{16}{24} \approx 0.667 = 66.7\%$

iii. $P(A \text{ and } B) = \dfrac{2}{24} \approx 0.083$

$= 8.3\%$

iv. $P(A \text{ or } B) = \dfrac{19}{24} \approx 0.791$

$= 79.1\%$

v. $P(B \text{ and } C) = \dfrac{14}{24} \approx 0.583$

$= 58.3\%$

vi. $P(B \text{ or } C) = \dfrac{21}{24} = 0.875$

$= 87.5\%$

STUDENT PAGE 56

Practice and Applications

16. The two-way table below shows data collected from a group of students in a science class who had been studying the relationship between eye dominance and hand dominance.

		Eye Dominance		
		Left	Right	Totals
Hand Dominance	Left	3	2	5
	Right	5	14	19
	Totals	8	16	24

Let *A* represent the set of students who are left-hand dominant.

Let *B* represent the set of students who are right-eye dominant.

Let *C* represent the set of students who are right-hand dominant.

a. Draw a Venn diagram that shows the relationship between left-hand and right-eye dominance. You can use *Activity Sheet 13*. On the diagram label all the sections and show the number of students in each circle, in the intersection of the two circles, and in the exterior of both circles.

Since *A* represents the set of students who are left-hand dominant, then the symbol $P(A)$ represents the probability of selecting a student who is left-hand dominant.

b. Use the descriptions for *A*, *B*, and *C* to describe each of the following and find the indicated probability.

i. $P(C)$

ii. $P(B)$

iii. $P(A \text{ and } B)$

iv. $P(A \text{ or } B)$

v. $P(B \text{ and } C)$

vi. $P(B \text{ or } C)$

STUDENT PAGE 57

17. a. i. 0.22

  ii. 0.65

  iii. 0.09

  iv. 0.22 + 0.65 − 0.09 = 0.78

  v. 0.57 + 0.21 = 0.78

  vi. 0.35 + 0.21 − 0.07 = 0.49

  vii. 0.21 + 0.22 = 0.43

b. Parts v and vii no values in the intersection; and parts iv and vi values in the intersection

---

17. A government agency was interested in how the descriptions of workers in the United States changed from 1985 to 1995. The table below shows the relative frequencies of new workers from 1985 to 1995.

Description of Workers

	White	Nonwhite	Immigrant	Totals
Men	0.15	0.07	0.13	0.35
Women	0.42	0.14	0.09	0.65
Totals	0.57	0.21	0.22	1.00

a. What is the probability that a randomly selected new worker is

  i. an immigrant?

  ii. a woman?

  iii. an immigrant and a woman?

  iv. an immigrant or a woman?

  v. white or nonwhite?

  vi. a man or nonwhite?

  vii. nonwhite or an immigrant?

b. For which of the questions in part a can the Addition Rule be used?

LESSON 7

Complementary Events

Materials: Lessons 6 and 7 Quiz (optional)
Technology: calculator
Pacing: $\frac{1}{2}$ class period

Overview

This lesson continues to formulate rules of probability. In this lesson, complementary events are defined and students are asked to find the probability of the complement of an event. Students are expected to use the Rule of Complementary Events to find that P(1 or more) = P(at least 1). This lesson also contains additional problems dealing with the AND and OR Rules.

Teaching Notes

Students may have difficulty with the idea that "at least 1" is the complement of 0. You may wish to write all the outcomes of rolling a die and asking students to find the complement of the following events.

1. rolling a 6
2. rolling a 1
3. rolling an even number
4. rolling a number greater than 1
5. rolling a number less than 6

For each event, have students list the given value in pencil and the complement in pen. All six numbers will be listed, but the different color will show the relationship between an event and its complement.

Follow-Up

You may wish to introduce the term "mutually exclusive events." Mutually exclusive events are events that cannot occur at the same time. $P(A \text{ and } B) = 0$ if A and B are mutually exclusive events.

STUDENT PAGE 58

Complementary Events

If 3 students in a class of 28 are absent, how many students are not absent?

If you toss a number cube and it does not land with a 5 up, what numbers might be showing?

If every member of your family went to a movie, how many stayed home?

OBJECTIVE

Find the probability of the complement of an event.

Each of the situations above asks you to find the *complement* of an event. In your class, the female students are the complement of the male students. In the alphabet, the vowels are the complement of the consonants.

INVESTIGATE

In this lesson, you will study a number of situations which involve events and their complements.

Discussion and Practice

1. A school administrator, interested in how many students were absent from an 8th-hour study hall, kept track of absences over the last 40 days of school. The table shows that there were 7 days when no students were absent, 8 days when 1 student was absent, and 4 days when 2 students were absent.

STUDENT PAGE 59

Solution Key

Discussion and Practice

1. **a.** 3 days

    b. 33 days; "at least 1" involves the total from 1 student to 10 students. Either add the number of days from 1 to 10; or subtract 7, the number of days not involved in the question, from 40.

    c. 9 days; "at least 5 students" involves the number of students absent from 5 to 10. Add the number of days when there were 5 to 10 students absent.

    d. 39 days; "fewer than 10 students" involves the outcomes 0 to 9. Either add these outcomes; or subtract 1 day, the number of days 10 students were absent, from the total of 40 days.

2. **a.** The complement is the number of days 1 or more students were absent.

    b. The complement is the number of days fewer than 1 student was absent, or the number of days no students were absent.

3. **a.** $\frac{7}{40} = 0.175 = 17.5\%$

    b. $\frac{33}{40} = 0.825 = 82.5\%$

    Note: Stress here that the probabilities requested are complementary events.

4. 1, or 100%

Number of Students Absent	Days
0	7
1	8
2	4
3	6
4	6
5	3
6	2
7	2
8	1
9	0
10	1
Total	40

a. For how many days were 5 students absent from study hall?

b. For how many days was at least 1 student absent from study hall? Explain how you found your answer.

c. For how many days were at least 5 students absent from study hall? Explain how you found your answer.

d. For how many days were fewer than 10 students absent from study hall?

When you found the number of days with fewer than 10 students absent from study hall, you were finding the complement of the number of days with 10 students absent. The *complement* of event *A* is the event that *A* does *not* occur, written "not *A*."

2. Consider the data given in Problem 1.

    a. If *A* represents the number of days that no students were absent from study hall, what is the complement of *A*?

    b. If *B* represents the number of days that at least 1 student was absent from study hall, what is the complement of *B*?

3. If the school administrator randomly selects one of the 40 days, what is the probability

    a. that no students were absent from study hall?

    b. that at least 1 student was absent from study hall?

4. What is the sum of the probabilities from Problems 3a and 3b?

STUDENT PAGE 60

5. **a.** The complement of event *A* is a person that does not have type A blood. This could also be described as a person who has type O, B, or AB blood.

    b. **i.** 0.40

       ii. 0.45

     iii. 0.40 + 0.45 = 0.85

     iv. 1 − (0.10 + 0.40) = 1 − 0.50 = 0.50

Rule of Complementary Events

$P(A) + P(\text{not } A) = 1$

$P(\text{not } A) = 1 - P(A)$

$P(A) = 1 - P(\text{not } A)$

The rule above can be used to find the probability that 1 or more students were absent, or the complement of no students absent.

$P(\text{one or more students absent}) = 1 - P(\text{no students absent})$

$$= 1 - \frac{7}{40}$$

$$= \frac{33}{40}$$

5. The following table gives the approximate relative frequency of Americans that have certain blood types.

Type	Relative Frequency
O	0.45
A	0.40
B	0.10
AB	0.05

Let *O* represent a person who has type O blood.

Let *A* represent a person who has type A blood.

Let *B* represent a person who has type B blood.

Let *AB* represent a person who has type AB blood.

a. What is the complement of *A*?

b. If a random American is chosen, find each of the following:

    i. $P(A)$

    ii. $P(O)$

   iii. $P(A \text{ or } O)$

   iv. $P(\text{not } B \text{ and not } A)$

STUDENT PAGE 61

6. **a.** The complement of *A* is the event that a student did not get 10 wrong.

b. The complement of *B* is the event that a student got fewer than 2 wrong.

c. *P*(2 or more wrong) = 1 − *P*(fewer than 2 wrong) = 1 − ($\frac{1}{30}$)

= $\frac{29}{30}$ ≈ 0.967

d. **i.** 0

ii. $\frac{5}{30}$ ≈ 0.167 = 16.7%

iii. $\frac{25}{30}$ ≈ 0.833 = 83.3%

iv. $\frac{13}{30}$ ≈ 0.433 = 43.3%

v. $\frac{1}{30}$ ≈ 0.033 = 3.3%

vi. $\frac{26}{30}$ ≈ 0.867 = 86.7%

e. "At least 1 wrong" indicates the events of 1 or more wrong. The complement of this event is "fewer than 1 wrong." As only 5 students had fewer than 1 wrong, the Rule of Complementary Events indicates

$1 - \frac{5}{30} = 1 - 0.167 = 0.833 = 83.3\%$.

6. Below is the tally of the number of wrong answers that a class of 30 students received on a mathematics test.

Number Wrong	Number of Students
1	/
2	//
3	/
4	////
5	ЖⱧ
6	////
7	///
8	/
9	/
10	
11	//
12	/

The total number of students is 30.

a. If *A* represents the event that a student got 10 answers wrong, what is the complement of *A*?

b. If *B* represents the event that a student got 2 or more answers wrong, what is the complement of *B*?

c. Use the Rule of Complementary Events to find *P*(2 or more answers wrong).

d. What is the probability that a randomly selected student

i. got 10 wrong?

ii. did not get any answers wrong?

iii. got at least 1 answer wrong?

iv. got 4, 5, or 6 answers wrong?

v. got at most 1 answer wrong?

vi. did not get exactly 4 answers wrong?

e. Explain how the Rule of Complementary Events could be used for part d iii.

STUDENT PAGE 62

Practice and Applications

7. **a.** The complement of B is that a person was not born in June.

b. There are 30 days in June; therefore, the probability of selecting a birthday not in June is

$$1 - \frac{30}{366} = \frac{336}{366} \approx 0.918 = 91.8\%.$$

c. As indicated in the solution to part b, the probability of selecting a birthday *not* in June is, by the Rule of Complementary Events, $1 - P$(selecting a birthday in June).

d. **i.** P(birthday in June)

$$+ \ P\text{(birthday in July)} = \frac{30}{366}$$

$$+ \frac{31}{366} = \frac{61}{366} \approx 0.167 = 16.7\%$$

ii. P(birthdays on the 31st of a month, including October) + P(birthday in October) − P(birthday on October 31)

$$= \frac{7}{366} + \frac{31}{366} - \frac{1}{366} = \frac{37}{366}$$

$$\approx 0.101 = 10.1\%$$

iii. P(not in September or November) $= 1 - [P$(birthday in September) + P(birthday in November)] $= 1 - (\frac{30}{366} + \frac{30}{366})$

$$= 1 - \frac{60}{366} = \frac{306}{366} \approx 0.836 = $$

83.6%

iv. P(birthday not June 20th) $=$ $1 - P$(birthday June 20th) $=$

$$1 - \frac{1}{366} = \frac{365}{366} \approx 0.997 = $$

99.7%

8. **a.** The complement is the event of completing high school or beyond.

b. P(did not complete high school)

$$= \frac{13,183}{28,528} \approx 0.462 = 46.2\%$$

c. Note: You may need to point out that the event "Completed High School" indicates that the subjects completed only high school. This number is not added,

---

Practice and Applications

7. In the late 1960s and the early 1970s, men were drafted into the U.S. Army by random selections of birthdays. Assume that each of the 366 birthdays has an equal chance of being chosen.

   a. If B represents the event that a person was born in June, what is the complement of B?

   b. Find the probability that a randomly selected birthday is not in June.

   c. Explain how the Rule of Complementary Events helps to answer part b.

   d. What is the probability that a randomly selected birthday is

      i. in June or July?

      ii. the 31st of a month or in October?

      iii. not in September or November?

      iv. not June 20th?

8. The table below contains Census data on the education attainment for adults more than 65 years old.

Education	Number (thousands of persons)
Did Not Complete High School	13,183
Completed High School	9,412
College, 1–3 Years	2,915
College, 4 or More Years	3,018
Total	28,528

   a. If E is the event that a person did not complete high school, what is the complement of E?

   b. Find the probability that a randomly selected person did not complete high school.

   c. Find the probability that a randomly selected adult more than 65 years old

      i. completed high school.

      ii. completed some college but less than 4 years of college.

      iii. completed high school or college.

---

or accumulated, into the other categories of college.

   i. This question does not ask "Completed High School Only"; so, the probability that an adult completed high school is: P(high school) + P(college, 1–3 years) + P(college, 4 or more) =

$$\frac{9412 + 2915 + 3018}{28,528} = \frac{15,345}{28,528}$$

$$\approx 0.538 = 53.8\%.$$

   ii. P(some college, less than 4)

$$= \frac{2915}{28,528} \approx 0.102 = 10.2\%$$

   iii. P(high school or college) $= P$(high school) + P(college, 1–3 years) + P(college, 4 or more) $= \dfrac{9412 + 2915 + 3018}{28,528}$

$$\approx 0.538 = 53.8\%$$

You may wish to show students this method:

$1 - P$(did not complete high school) $= 1 - \dfrac{13,183}{28,528} = \dfrac{15,345}{28,528}$

$$\approx 0.538 = 53.8\%$$

9.
  a. 0.55 = 55%

  b. 0.84 = 84%

  c. 0.54 = 54%

  d. P(knew answer to Question 1) + P(knew answer to Question 2) − P(knew answers to Questions 1 and 2) = 0.55 + 0.84 − 0.54 = 0.85 = 85%

  e. 0.15 = 15%

  f. The complement of A is a person who knew what city will host the 1996 Summer Olympics.

STUDENT PAGE 63

9. During the 1992 Winter Olympics, a random survey asked 2600 American adults these questions.

Question 1: What city hosted the 1992 winter Olympics?

Question 2: What city will host the 1996 Summer Olympics?

The table below shows the relative frequencies for the results of the answers to these two questions.

		Question 1	
		Knew	Did Not Know
Question 2	Knew	0.54	0.30
	Did Not Know	0.01	0.15

If a randomly selected American adult had been selected, what is an estimate of the probability that

  a. he or she knew the answer to Question 1?

  b. he or she knew the answer to Question 2?

  c. he or she knew the answer to both questions?

  d. he or she knew the answer to Question 1 or Question 2?

  e. he or she did not know the answer to either question?

  f. If A represents the event that a person did not know what city would host the 1996 Summer Olympics, what is the complement of A?

Conditional Probability

Materials: none
Technology: calculator
Pacing: $1\frac{1}{2}$ class periods

Overview

This lesson builds on interpreting data presented in a two-way table and forms the basis for the remaining lessons in the module. Problems 1–3 ask students to complete a table that has some cell and marginal totals missing. These problems allow students to review Lesson 6, where tables were first introduced, and provides the opportunity for them to preview the rest of the lesson. Problems 4–9 introduce valuable vocabulary related to two-way tables. Students are asked to find marginal totals, marginal relative frequencies, joint frequency, joint relative frequency, and conditional relative frequency from data in a two-way table. The lesson concludes by having students relate conditional relative frequencies to conditional probability.

Teaching Notes

The vocabulary in this lesson can be confusing to students. You may wish to present a two-way table with the marginal and joint frequencies labeled with different colors. Below is another example that you may wish to use with students. To represent two colors, joint frequencies are shown in regular type, while marginal frequencies are shown in bold type.

| | Do You Participate in a Sport? | | |
	Yes	No	Totals
Do You Play a Musical Instrument? Yes	5	4	**9**
No	7	9	**16**
Totals	**12**	**13**	**25**

Students should be able to state clearly what each of the values in the table represents before changing the values to relative frequencies. The next step is to present the table showing the data as relative frequencies. Point out to students that each of the values was found by dividing each joint frequency by the grand total. Here again, joint relative frequencies are shown in regular type (one color), while marginal relative frequencies are shown in bold type (another color).

| | Do You Participate in a Sport? | | |
	Yes	No	Relative Frequencies
Do You Play a Musical Instrument? Yes	0.2	0.16	**0.36**
No	0.28	0.36	**0.64**
Totals	**0.48**	**0.52**	**1.00**

As before, students should be able to state clearly what each of the values in the table represents and how the values were calculated.

When discussing conditional relative frequency, you may wish to show students the entire table and then cover up the row or the column that you are not using. For example, among students who said they play a musical instrument, what is the conditional relative frequency of students who said they participate in a sport? You would cover up the row labeled "No" for playing a musical instrument.

Do You Play a Musical Instrument?		Yes	No	Total
	Yes	5	4	9

The conditional relative frequency is $\frac{5}{9} \approx 0.555$.

Follow-Up

Collect data from students on the number of letters in their first names and build a two-way table similar to the one shown in Problem 11. After the table is completed, ask some probability questions pertaining to the data. You may wish to use these data for a simulation. For the question "What is the probability that the name of a randomly selected student has more than 6 letters?" you could randomly select a student and tally whether or not his or her name has more than 6 letters. Replacing this name and repeating the experiment many times would reinforce the ideas from Lessons 1 and 2. For the question "Considering only females, what is the probability that her name has more than 6 letters?" you would randomly select from only the females in class and tally the results.

STUDENT PAGE 64

LESSON 8

Conditional Probability

When you survey people with two questions, does a "yes" response to one question help you to predict the answer to the other question?

How do basic rules of probability help you to make predictions?

OBJECTIVES

Construct and interpret relative frequencies from columns or from rows of a table.

Interpret column or row relative frequencies as conditional probabilities.

In the preceding lessons, you have investigated several of the basic rules of probability. Many of the examples involved the examination of data, summarized in a two-way table, that resulted from asking people two questions. Is a person who has seen one popular movie likely to have seen another? Is a person who has right-hand dominance likely to have right-eye dominance?

INVESTIGATE

In this lesson, you will use tables to help you understand probability and make predictions.

Discussion and Practice

According to the United States Bureau of the Census, the number of family households in the country classified according to household head and status of children for 1993 is as shown in the following table.

STUDENT PAGE 65

Solution Key

Discussion and Practice

1. a. 28.4 million, or 28,400,000, married-couple households have no children under the age of 18.

b. See table below.

c. 24.7 million married couples have children under the age of 18.

d. 1.7 million households headed by a male have no children under the age of 18.

e. 68 million

2. a. $\frac{11.9}{68} \approx 0.175 = 17.5\%$

b. $\frac{33.2}{68} \approx 0.488 = 48.8\%$

c. $\frac{1.3}{68} \approx 0.019 = 1.9\%$

3. a. This question can be answered. 50 students were surveyed. If 36 indicated they eat breakfast at least 3 times per week, then 14 would indicate they do not.

	Married-Couple Household Head (millions)	Male Household Head (millions)	Female Household Head (millions)	Totals
No Children Under 18	28.4		4.7	
Children Under 18	24.7	1.3	7.2	
Totals	53.1	3.0		

Source: United States Bureau of the Census

1. Use the data from the table above for these problems.

 a. What does the number 28.4 million represent?

 b. Copy and complete the table.

 c. How many households headed by a married couple have children under 18?

 d. How many households headed by a male have no children under 18?

 e. What is the total number of family households in the United States?

2. Use the data from the table above for these problems. If the A.C. Nielsen company, which conducts TV ratings surveys, randomly selected a household, what is the probability that the household

 a. is headed by a female?

 b. has children under 18?

 c. is headed by a male and has children under 18?

3. In the Rufus King High School survey introduced in Lesson 5, two of the many questions that were asked were:

 Do you consider your diet to be healthy?

 Do you eat breakfast at least 3 times a week?

 Among the 50 students in a random sample from the school, 20 answered "yes" to the first question and 36 answered "yes" to the second question.

 Several questions are listed below. Which of the questions can be answered from the survey data given and which cannot? If a question cannot be answered, explain why not.

 a. How many students do not eat breakfast at least 3 times a week?

	Married-Couple Household Head (millions)	Male Household Head (millions)	Female Household Head (millions)	Totals
No Children Under 18	28.4	1.7	4.7	34.8
Children Under 18	24.7	1.3	7.2	33.2
Totals	53.1	3.0	11.9	68

STUDENT PAGE 66

b. This question cannot be answered from the given data. It is necessary to re-examine the surveys and determine each student's answers to the two questions.

c. This question can be answered, as it is the complement of the results to the question "Do you consider your diet to be healthy?" As 50 students were surveyed and 20 indicated yes to the first question, 30 students would consider their diet to not be healthy.

d. The probability of selecting a student from the sample of 50 who eat breakfast at least 3 times a week is $\frac{36}{50}$, or 72%. Use this relative frequency as a way to estimate a similar response from all Rufus King students:
$1200 \times 0.72 = 864$ students

4. **a.** See table below.

b. This question is included so students discuss the process of completing the table values.

c. 14 students answered "no" to the breakfast question.

d. 30 students answered "no" to the healthy-diet question.

e. 11 students answered "no" to both questions.

b. How many students eat breakfast at least 3 times a week and consider their diet healthy?

c. How many students do not consider their diet healthy?

d. If the total enrollment in the school is 1200 students, approximately how many students in the school would you expect to eat breakfast at least 3 times a week?

Two-Way Tables

When analyzing the Rufus King data, it was not possible to answer the question "How many surveyed students eat breakfast at least 3 times a week and consider their diet to be healthy?" To answer a question like this implies that there are responses for each student on both parts of the question. If you are interested in investigating a relationship between the two variables, eating breakfast and perceptions of healthy diet, you must keep track of how each individual responded to both parts. As you saw in Lesson 6, one way to do this is to record the data in a two-way table.

4. The two-way table below shows the results of the Rufus King survey. The 17 represents the number of students who answered "yes" to both questions.

		Do You Think Your Diet Is Healthy?		
		Yes	No	Totals
Do You Eat Breakfast at Least 3 Times a Week?	Yes	17		36
	No			
	Totals	20		50

a. Copy and complete the table above.

b. Compare your results to those of others in your group.

c. How many students answered "no" to the breakfast question?

d. How many students answered "no" to the healthy-diet question?

e. How many students answered "no" to both questions?

Totals for each row and column are called ***marginal totals***. The 36 in the first row of the table above is the marginal total of the students who said "yes" to the breakfast question; 20 is the marginal total of the number of students who answered "yes" to the diet question. To help in an analysis of the data, it is

Do You Eat Breakfast at Least 3 Times a Week?

	Do You Think Your Diet Is Healthy?		
	Yes	No	Totals
Yes	17	19	36
No	3	11	14
Totals	20	30	50

STUDENT PAGE 67

5. a. See table below.

b. 72% of the students in the sample eat breakfast at least 3 times a week.

c. 0.72 = 72%

d. The marginal relative frequency indicates that 0.60, or 60%, of the students in the sample do not think their diet is healthy.

e. 40% of 1200 = 0.40 × 1200 = 480 students

6. a. 3 students

b. 19 students

sometimes helpful to convert the marginal totals to marginal relative frequencies or percents.

5. Use the table from the Rufus King survey.

   a. Copy and complete the table below by converting the marginal totals to marginal relative frequencies.

| | Do You Think Your Diet Is Healthy? | | |
	Yes	No	Totals
Yes			$\frac{36}{50} = 0.72$
No			
Totals	$\frac{20}{50} = 0.40$		$\frac{50}{50} = 1.00$

Do You Eat Breakfast at Least 3 Times a Week?

   b. What does the marginal relative frequency 0.72 tell you?

   c. What is the probability that a student randomly selected from the sample of 50 students will answer "yes" to the breakfast question?

   d. What does the marginal relative frequency under the *No* column for the diet question tell us?

   e. Of the 1200 students in the school, approximately how many would you expect to consider their diet healthy?

The values in the table provide information on the *joint* behavior of students in response to the two questions. The value 17, which is the number of students who answered "yes" to both questions, is called a *joint frequency*.

6. Consider the marginal relative frequencies from the table above.

   a. What is the joint frequency of the students who answered "yes" to the diet question and "no" to the breakfast question?

   b. What is the joint frequency of the students who answered "no" to the diet question and "yes" to the breakfast question?

To convert the data to *joint relative frequencies,* divide each joint frequency by the total number. The joint relative frequency of the students who answered "yes" to both questions is $\frac{17}{50} = 0.34$.

Do You Think Your Diet Is Healthy?

	Yes	No	Totals
Yes	17	19	$\frac{36}{50} = 0.72$
No	3	11	$\frac{14}{50} = 0.28$
Totals	$\frac{20}{50} = 0.40$	$\frac{30}{50} = 0.60$	$\frac{50}{50} = 1.00$

Do You Eat Breakfast at Least 3 Times a Week?

STUDENT PAGE 68

7. **a.** See table below.

b. This question is included so students discuss how they obtained the answers for the table.

c. 0.34 indicates that 34% of the students in the sample, or 17 students, answered "yes" to both questions. This joint relative frequency could also be used to estimate the probability of selecting a student from the school who thinks his or her diet is healthy and eats breakfast at least 3 times a week.

d. This joint relative frequency indicates that 0.06, or 6%, of the sample answered "yes" to the diet question and "no" to the breakfast question. This joint relative frequency could also be used to estimate the probability of selecting a student from the school that answered the two questions in the same way.

e. 0.06 = 6%

8. **a.** $\frac{17}{36}$, or as the previous explanation indicated, $\approx 0.47 = 47\%$

b. $\frac{19}{36} \approx 0.53 = 53\%$

c. No; slightly more than half the students who eat breakfast at least 3 times a week indicated their diet is not healthy.

---

7. Use the table in Problem 4.

   a. Convert the joint frequencies to joint relative frequencies.

   b. Compare your results to those of others in your group.

   c. What does the joint relative frequency of $\frac{17}{50}$, or 0.34, represent?

   d. What does the joint relative frequency in the cell represented by "yes" on the diet question and "no" on the breakfast question represent?

   e. What is the probability that a randomly selected student from among the 50 students selected in the sample will answer the diet question "yes" and the breakfast question "no"?

If a student said that he or she ate breakfast at least 3 times a week, is this student more likely to say that he or she has a healthy diet than an unhealthy diet? To help answer this question, you will need to study the first row of the table—that is, the row that represents the students that said "yes" to the breakfast question.

	Do You Think Your Diet Is Healthy?		
	Yes	**No**	**Totals**
Do You Eat Breakfast at Least 3 Times a Week? Yes	17	19	36

From this total of 36 students, 17, or 47%, of the students said "yes" to the diet question. The 47% is called the *conditional relative frequency*.

8. Use the data above to answer the following questions.

   a. Among the students known to eat breakfast at least 3 times a week, what is the conditional relative frequency of students who think their diet is healthy?

   b. Among the students known to eat breakfast at least 3 times a week, what is the conditional relative frequency of students who think their diet is not healthy?

   c. If a student said that she ate breakfast at least 3 times a week, is she more likely to say that she has a healthy diet than an unhealthy diet?

Do You Think Your Diet Is Healthy?

		Yes	**No**	**Totals**
Do You Eat Breakfast at Least 3 Times a Week?	**Yes**	$\frac{17}{50} = 0.34$	$\frac{19}{50} = 0.38$	$\frac{36}{50} = 0.72$
	No	$\frac{3}{50} = 0.06$	$\frac{11}{50} = 0.22$	$\frac{14}{50} = 0.28$
	Totals	$\frac{20}{50} = 0.40$	$\frac{30}{50} = 0.60$	$\frac{50}{50} = 1.00$

(8) d. Using the marginal relative frequency of 0.72, 72% of the 1200 is used to estimate the number of students who eat breakfast:
$0.72 \times 1200 = 864$ students
Of these 864 students, an estimate of 47% is used to determine the number of students who think their diet is healthy, that is,
$0.47 \times 864 = 406$ students.

9. a. $\frac{3}{14} \approx 0.214 = 21.4\%$

b. $\frac{11}{14} \approx 0.786 = 78.6\%$

c. $\frac{14}{50}$, or 28%, is used to estimate the number of students who do not eat breakfast at least 3 times a week. With the given school population of 1200 students, $0.28 \times 1200 = 336$ students do not eat breakfast at least 3 times a week. Of these 336 students, you would expect 0.214, or 21.4%, to think their diet is healthy. This indicates that $0.214 \times 336 = 72$ students who do not eat breakfast 3 times a week would think their diet is healthy.

STUDENT PAGE 69

d. Of the 1200 students in the school, approximately how many would you expect to eat breakfast at least 3 times a week? Among these breakfast-eaters, approximately how many would you expect to think their diet is healthy?

The conditional relative frequency can also be thought of as a *conditional probability*.

9. Consider those students who said "no" to the breakfast question and how they answered the diet question.

	Do You Think Your Diet Is Healthy?		
Do You Eat Breakfast at Least 3 Times a Week?	**Yes**	**No**	**Totals**
No	3	11	14

a. If a student is randomly selected from those who are not regular breakfast-eaters, what is the conditional probability that this student will say that his or her diet is healthy?

b. If a student is randomly selected from those who are not regular breakfast-eaters, what is the conditional probability that this student will say that his or her diet is not healthy?

c. Of the 1200 students in the school, approximately how many would you expect to not eat breakfast at least 3 times a week? Among those who don't eat breakfast, approximately how many would you expect to think their diet is healthy?

Summary

Data from variables that classify items into categories can be summarized by recording the frequencies for each category. If two categorical variables are to be compared, the frequencies can be arranged in a two-way table. A *joint relative frequency* or probability is the ratio of a cell frequency to the overall number of times classified in the table. A *marginal relative frequency* or probability is the ratio of a row or column total to the overall frequency of items for the whole table. Marginal data provide information on either the row or column variable by itself. A *conditional relative frequency* or probability is the ratio of a cell frequency to either its row total or its column total.

STUDENT PAGE 70

Practice and Applications

10. a. See table below.

b. Circles should be placed around the 32, 18, 30, and 20.

c. See table below.

d. 0.60 = 60%

e. 0.36 = 36%

f. $\frac{18}{32} \approx 0.563 = 56.3\%$

g. $\frac{14}{32} \approx 0.437 = 43.7\%$

h. 0.64 = 64%; as a result, you would expect 0.64 × 1200 = 768 students to indicate that they participate in sports. You would expect 56.3% of this group to like school, that is, 0.563 × 768 = 432 students.

Practice and Applications

10. The Rufus King High School survey also included the following two questions:

Do you like school?

Do you participate in a sport at school?

a. Copy and complete the table below.

	Do You Like School?		
	Yes	No	Totals
Yes		14	32
No	12		
Totals			50

(Do You Participate in a Sport?)

b. In your table, circle the values that represent the marginal totals.

c. Construct a table showing the data converted to marginal and joint relative frequencies.

d. What is the approximate probability that a randomly selected student will answer "yes" to the school question?

e. What is the approximate probability that a randomly selected student will answer "yes" to both questions?

f. Consider only the students who said that they have participated in a sport. What is the conditional relative frequency of those students who like school within this group?

g. Consider only the students who said that they have participated in a sport. What is the conditional relative frequency of those students who do not like school within this group?

h. Of 1200 students in the school, approximately how many would you expect to have participated in a sport? Among those who participated, approximately how many would you expect to like school?

i. Consider only the students who said that they do not like school. What is the conditional relative frequency of the students who do not participate in sports within this group?

a.

Do You Like School?

Do You Participate in a Sport?	Yes	No	Totals
Yes	18	14	32
No	12	6	18
Totals	30	20	50

c.

Do You Like School?

Do You Participate in a Sport?	Yes	No	Totals
Yes	$\frac{18}{50} = 0.36$	$\frac{14}{50} = 0.28$	$\frac{32}{50} = 0.64$
No	$\frac{12}{50} = 0.24$	$\frac{6}{50} = 0.12$	$\frac{18}{50} = 0.36$
Totals	$\frac{30}{50} = 0.60$	$\frac{20}{50} = 0.40$	$\frac{50}{50} = 1.00$

STUDENT PAGE 71

(10) i. $\frac{6}{20} = 0.3 = 30\%$

j. $\frac{12}{30} = 0.4 = 40\%$

11. a. See table below.

b. $\frac{14}{26} \approx 0.538 = 53.8\%$

c. $\frac{8}{16} = 0.5 = 50\%$

d. $\frac{4}{26} \approx 0.154 = 15.4\%$

e. $\frac{3}{26} \approx 0.115 = 11.5\%$

f. $\frac{3}{10} = 0.3 = 30\%$

g. $\frac{3}{16} \approx 0.188 = 18.8\%$

h. $\frac{5}{8} = 0.625 = 62.5\%$

j. Consider only the students who said that they like school. What is the conditional relative frequency of those students who do not participate in sports within this group?

11. The following data were collected from a high-school mathematics class.

Number of Letters in First Name

	3 or Fewer	4 to 6	More than 6	Totals
Male	1	6	3	
Female	3	8	5	
Totals				

a. Copy the table and fill in the marginal totals for each row and column.

b. What percent of students have 4 to 6 letters in their first names?

c. What percent of females have 4 to 6 letters in their names?

d. If you were to choose a student from this class at random, what is the probability that the student would have 3 or fewer letters in his or her name?

e. If you were to choose a student from this class at random, what is the probability that the student would be male and have more than 6 letters in his name?

f. If the person randomly chosen from this class is a male, what is the probability that this male will have more than 6 letters in his name?

g. If a female is chosen from this class, what is the probability that she has 3 or fewer letters in her name?

h. If the name that is chosen has 6 or more letters, what is the probability that the person is female?

Number of Letters in First Name

	3 or Fewer	4 to 6	More than 6	Totals
Male	1	6	3	10
Female	3	8	5	16
Totals	4	14	8	26

Assessment for Unit III

Materials: none
Technology: calculator (optional)
Pacing: 1 class period or homework

Overview

This assessment enables you to evaluate students'
understanding of Lessons 5–8. Problem 1 covers
material from Lessons 5–7. Problems 2 and 3 cover
material from Lessons 6 and 7. Problems 4, 5, 7, and
8 cover material from Lesson 8, while Problem 6 covers material from Lesson 5.

Teaching Notes

As your students work through the problems, stress
the need to have them clearly show their work and
their method of solving the problem.

If you stressed the use of Venn diagrams, you may
wish to have your students display Venn diagrams for
Problem 7.

Follow-Up

The four probability questions that students write for
Problem 8b could be compiled, and the class could discuss the different type of questions that were written.

Solution Key

1. **a.** 0.16 = 16%
   b. 1 – 0.16 = 0.84 = 84%
   c. 4% + 9% = 13%
   d. 1 – 0.13 = 0.87 = 87%
   e. 16% + 4% + 9% = 29%

STUDENT PAGE 72

ASSESSMENT

Assessment for Unit III

OBJECTIVE

Apply the concepts of compound events, complementary events, and conditional probability.

1. In a nationwide survey, 1250 adults were asked this question: "Where in the world are the safest cars produced?" The following graph summaries their responses.

Safest Car Survey

Country	Percent
U.S.	52%
Europe	18%
Japan	9%
No Difference	4%
Don't Know	16%

What is an estimate for the probability that a randomly chosen adult will answer that the safest auto is produced

a. in the U.S.? **b.** not in the U.S.?

c. in Europe or in Japan?

d. not in Europe or not in Japan?

e. in the U.S. or in Europe or in Japan?

2. The following table shows the results of a Channel One News survey. The table shows the number of middle- and high-school students who say guilty teens deserve corporal punishment for certain crimes.

	Deserve Corporal Punishment		
	Yes	No	Totals
Middle/Junior High School	401	381	
High School	464	304	
Totals			

STUDENT PAGE 73

2. a. 768 students
b. 865 students
c. i.

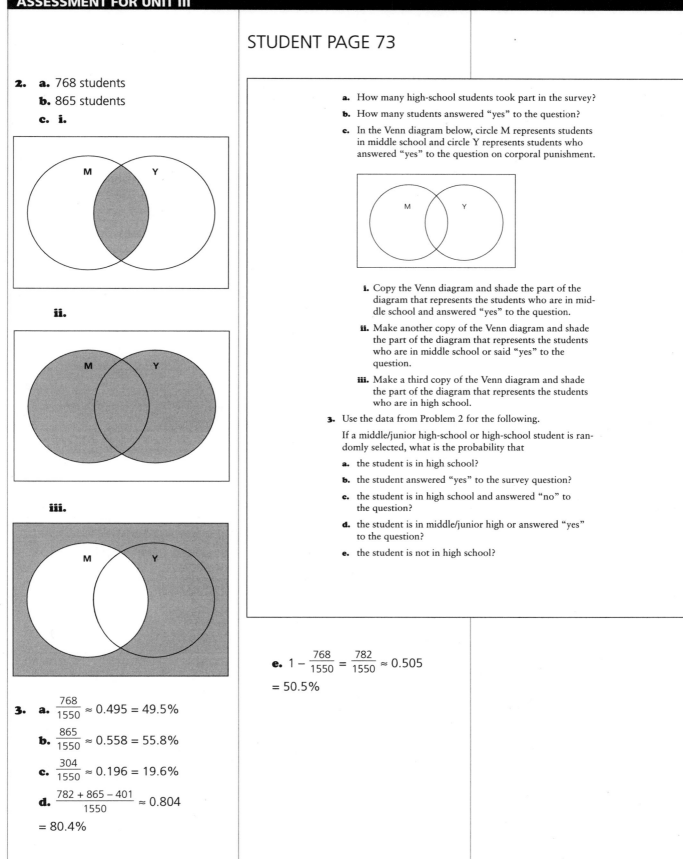

ii.

iii.

3. a. $\frac{768}{1550} \approx 0.495 = 49.5\%$

b. $\frac{865}{1550} \approx 0.558 = 55.8\%$

c. $\frac{304}{1550} \approx 0.196 = 19.6\%$

d. $\frac{782 + 865 - 401}{1550} \approx 0.804$
$= 80.4\%$

a. How many high-school students took part in the survey?

b. How many students answered "yes" to the question?

c. In the Venn diagram below, circle M represents students in middle school and circle Y represents students who answered "yes" to the question on corporal punishment.

i. Copy the Venn diagram and shade the part of the diagram that represents the students who are in middle school and answered "yes" to the question.

ii. Make another copy of the Venn diagram and shade the part of the diagram that represents the students who are in middle school or said "yes" to the question.

iii. Make a third copy of the Venn diagram and shade the part of the diagram that represents the students who are in high school.

3. Use the data from Problem 2 for the following.

If a middle/junior high-school or high-school student is randomly selected, what is the probability that

a. the student is in high school?

b. the student answered "yes" to the survey question?

c. the student is in high school and answered "no" to the question?

d. the student is in middle/junior high or answered "yes" to the question?

e. the student is not in high school?

e. $1 - \frac{768}{1550} = \frac{782}{1550} \approx 0.505$
$= 50.5\%$

STUDENT PAGE 74

4. a. $\frac{464}{768} \approx 0.604 = 60.4\%$

b. $\frac{464}{865} \approx 0.536 = 53.6\%$

5. a. 100% − 63% = 37%

b. No; the two descriptions— would rather use a computer than read a book and would rather use a computer than watch TV—are derived from two separate survey questions. It would be necessary to examine both responses of each person to determine the requested probability.

6. a. i. 13.1%

ii. 7.9% + 18.6% = 26.5%

---

4. Use the data from Problem 2 to answer the following questions.

  a. Consider only the high-school students. What is the probability that a randomly selected high-school student answered "yes" to using corporal punishment?

  b. What is the probability that a randomly selected student is in high school if you know that the person selected the answer "yes" to using corporal punishment?

5. The National Computing Survey of 2800 adults and children reported that 63% of the children aged 11–17 would rather use a computer than read a book and 59% would rather use a computer than watch TV.

  a. If a person aged 11–17 is randomly selected, what is the probability that the person would say that he or she would rather read a book than use a computer?

  b. Can you determine the probability of a person aged 11–17 who would rather use a computer than read a book and rather use a computer than watch TV? Explain your answer.

6. The table below contains information on the educational background for adults aged 25 to 34 in 1993. This data is part of the Census information gathered by the U.S. government.

Education	Percent of Population
Did Not Complete High School	13.1%
Completed High School	35.9%
Some College, No Degree	19.2%
Associate or Vocational Degree	7.9%
Bachelor's Degree	18.6%
Advanced Degree	5.2%

Source: *Statistical Abstract of the United States,* 1994

  a. Find the probability that a randomly selected adult aged 25 to 34

   i. did not complete high school by 1993.

   ii. received an associate, vocational, or bachelor's degree by 1993.

STUDENT PAGE 75

b. 99.9%; the significance of this value is that it covers all of the population surveyed. The difference between 99.9% and 100% could be due to rounding when calculating a percent.

c. The category "Completed High School" means "Completed Only High School." As a result, an estimate of the total number who completed high school would involve adding the percents of the categories "Completed High School," "Some College, No Degree," "Associate or Vocational Degree," "Bachelor's Degree," and "Advanced Degree." This can most easily be determined by subtracting the percent in the category "Did Not Complete High School" from 100%: 100% − 13.1% = 86.9%. Finally, 86.9% of 1000 = 0.869 × 1000 = 869 adults.

7. a. See table below.
    b. 0.264 = 26.4%
    c. 0.63 = 63%
    d. 0.20 = 20%
    e. 0.405 = 40.5%
    f. 0

---

b. Find the sum of the relative frequencies for each of the five categories. What is the significance of this value?

c. Suppose the Gallup Organization polled 1000 randomly selected adults aged 25 to 34. How many would you expect to say that they completed high school?

7. Prior to the 1996 Summer Olympic Games in Atlanta, the Atlanta Committee for the Olympic Games surveyed 606 Americans and asked how much interest they had in going to Atlanta for the Olympic games. The results are shown below.

Amount of Interest

Age Group	Very Interested	Somewhat Interested	Not Interested
18–34	160	96	64
35–54	109	87	90

a. Find the marginal totals and the marginal relative frequencies.

b. What is the joint relative frequency that a person from this survey is 18–34 years old and says that he or she is very interested in going to Atlanta for the Olympics?

c. What is the probability that a person from this survey is 35–54 years old or says that he or she is somewhat interested in going to Atlanta for the Olympics?

d. If it is known that the person surveyed is aged 18–34, what is the probability that the person has no interest in going to Atlanta for the Olympics?

e. If it is known that the person surveyed is very interested in going to Atlanta for the Olympics, what is the probability that the person is aged 35–54?

f. If it known that the person surveyed is aged 18–34, what is the probability that he or she is aged 35–54?

Amount of Interest

Age Group		Very Interested	Somewhat Interested	Not Interested	Totals
	18–34	$\frac{160}{606} \approx 0.264$	$\frac{96}{606} \approx 0.158$	$\frac{64}{606} \approx 0.106$	$\frac{320}{606} \approx 0.528$
Age Group	35–54	$\frac{109}{606} \approx 0.180$	$\frac{87}{606} \approx 0.144$	$\frac{90}{606} \approx 0.148$	$\frac{286}{606} \approx 0.472$
	Totals	$\frac{269}{606} \approx 0.444$	$\frac{183}{606} \approx 0.302$	$\frac{154}{606} \approx 0.254$	$\frac{606}{606} = 1.00$

STUDENT PAGE 76

8. **a.** See table below.

b. Answers will vary.

Two examples involving joint relative frequencies follow.

What is the probability that a person picked at random from the sample would indicate he or she has enough leisure time but would work fewer hours for less pay?

What is the probability of selecting a person who is willing to work fewer hours for pay but does not think he or she has enough leisure time?

Two examples involving conditional relative frequencies follow.

A person who is willing to work fewer hours for less pay was asked if he or she has enough leisure time. What is the probability that this person will answer "yes"?

If a person indicated he or she has enough leisure time, what is the probability that this person is not willing to work fewer hours for less pay?

8. A random survey of 1000 adults found that 48% said that they do have enough leisure time. Only 14% of the 1000 adults said they would work fewer hours for less pay. The survey also found that only 5% said that they have enough leisure time and would be willing to work fewer hours for less pay.

a. Copy and complete the two-way table with the joint frequencies.

	Do You Have Enough Leisure Time?		
Would You Work Fewer Hours for Less Pay?		Yes	No	Totals
	Yes			
	No			
	Totals			1000

b. Write four probability questions and their answers involving the information in your two-way table. At least one question should involve the joint frequencies and at least one question should involve conditional relative frequency.

	Do You Have Enough Leisure Time?		
Would You Work Fewer Hours for Less Pay?		**Yes**	**No**	**Totals**
	Yes	50	90	140
	No	430	430	860
	Totals	480	520	1000

Understanding Association

LESSON 9

Association

Materials: none
Technology: graphing calculator or spreadsheet program (optional)
Pacing: 1 class period

Overview

This lesson builds on the ideas from Lesson 8. Students use conditional relative frequency to develop an understanding about association between two variables. The lesson begins by presenting the results of a survey and using the results to review conditional probability. A bar graph of the conditional relative frequencies is shown, and students are asked if there seems to be a relationship between the two variables. Problems 6–10 use data collected from the students concerning hand dominance and eye dominance. Students are asked to decide if there is an association between hand dominance and eye dominance; that is, if a person is right-handed can you predict which eye will be dominant? The problems present examples of two-way tables that illustrate different degrees of association. Students compare the class results to these different examples to help make a decision concerning the association between eye dominance and hand dominance.

Teaching Notes

Problems 1–5 review previous ideas on conditional relative frequency. This lesson builds on those ideas and presents a bar graph of the conditional relative frequencies. Students should understand how the bar graph was constructed. The bar graphs can be constructed using a spreadsheet but students should first have the opportunity of constructing these graphs by hand. Problems 6–10 use data collected on hand dominance and eye dominance. Some students may have difficulty deciding which eye is dominant. Remind the students that they need to focus on something at a distance and if the object does not move then the open eye is the dominant eye. The bar graphs that students construct give them a visual picture of the conditional probabilities. In this lesson, students make a decision about association based

on the bar graphs and the difference in the conditional relative frequencies. Students should develop a sense of how a bar graph showing association would look, as well as how one showing no association would look. Students could use the following data to complete tables that show association and no association.

A table showing association:

		Do You Participate in a Sport?		
		Yes	No	Totals
Do You Play a Musical Instrument?	**Yes**	9	0	9
	No	3	13	16
	Totals	12	13	25

A table showing no association:

		Do You Participate in a Sport?		
		Yes	No	Totals
Do You Play a Musical Instrument?	**Yes**	4	5	9
	No	8	8	16
	Totals	12	13	25

Students will need to be reminded that association means that knowing how a person responded to one question helps them predict how that person will respond to another question. The next two lessons will give students another method to help them determine whether or not there is an association between two variables.

Follow-Up

Use examples from previous lessons and have students decide if there is an association between the variables.

Solution Key

Discussion and Practice

1.

Too Much Violence?

	Yes	No	Totals
Men	176	224	400
Women	320	80	400
Totals	496	304	800

2. $\dfrac{496}{800} = 0.62 = 62\%$

STUDENT PAGE 79

LESSON 9

Association

If you are right-handed, are you also right-eyed?

Do people who never smoked cigarettes think that smoking reduces stress?

In this lesson you will investigate how to answer questions that involve relationships, or *association*, between two variables.

OBJECTIVE

Understand how conditional probability can be used to measure association between two variables.

INVESTIGATE

Violence on TV and in Movies

Have you ever walked out of a movie or turned off a TV show because of the violence in the show? Is there a relationship between how men and women answer this question?

Discussion and Practice

In September, 1996, the American Medical Association released the results of a survey concerning violence in TV shows, movies, music lyrics, and computer games. The survey included a randomly selected nationwide sample of 800 registered voters. The results of one question on the survey are shown below.

Question: Have you ever walked out of a movie or turned off a TV show because of the violence in the show?

	Yes	No
Men	176	224
Women	320	80

1. Make a copy of the table above and include the marginal totals.

2. What is the relative frequency of people in the survey that answered "yes" to the question?

3. The conditional relative frequency of the men who answered "yes" is $\frac{176}{400} = 0.44 = 44\%$. The conditional relative frequency of the men who answered "no" is $\frac{224}{400} = 0.56 = 56\%$.

4. The conditional relative frequency of the women who answered "yes" is $\frac{320}{400} = 0.80 = 80\%$. The conditional relative frequency of the women who answered "no" is $\frac{80}{400} = 0.20 = 20\%$.

5. The conditional relative frequencies indicate a major difference in the response to this question based on the sex of a person. Women indicated they would respond "yes" to this question by approximately a 2 to 1 margin in comparison to men; that is, 2 women to nearly 1 man would answer "yes." Note how the bar graph indicates the disproportional height of the "yes" bars to "no" bars when comparing the men to the women.

STUDENT PAGE 80

3. Consider only the men in the survey. What is the conditional relative frequency of men who answered "yes"? What is the conditional relative frequency of men who answered "no"?

4. Consider only the women in the survey. What is the conditional relative frequency of women who answered "yes"? What is the conditional relative frequency of women who answered "no"?

A bar graph of the conditional relative frequencies can be used to decide if there is a relationship between the sex of the person and how the person answered the question about leaving a show because of violent content.

The bar graph below shows the distribution of the conditional relative frequencies.

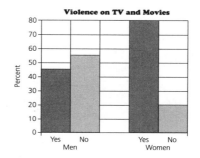

Violence on TV and Movies

5. Use the conditional relative frequencies and the bar graph to comment on whether there seems to be a relationship between the sex of a person and whether or not he or she would leave a show because of violent content. In other words, do the conditional relative frequencies seem to differ between men and women?

Left-Handed, Left-Eyed?

Is there a relationship between hand dominance and eye dominance? Check your eye dominance by making a circle with your thumb and forefinger and focusing, with both eyes open, on an object on the wall of the classroom. Close one eye and see if the object appears to move. If it does not move, the open eye is dominant.

6. Answers for the table and the accompanying questions will vary according to the data collected by your class. The table requires students to clearly indicate a right or left response to the two items. It is possible some students will have difficulty selecting a specific response, as they may notice a movement with either the left or the right eye; or, they may be those rare individuals who are comfortable with either hand. If possible, request each student select a "best" response to the items. If this is not possible, then do not include their responses in this sample. In other words, simply do not count the students who are clearly comfortable with both hands. The goal of this problem is to be able to direct attention to the "yes" and "no" responses for each question.

a. Using the data collected by the class, find the quotient $\frac{d}{c+d}$.

b. Using the data collected by the class, find the quotient $\frac{c}{c+d}$.

c. Evaluate the data according to the criteria indicated in the question. Most spreadsheet applications will produce this graph from the table. If using a spreadsheet application, highlight the cells including the labels; however, do not highlight the cells representing the totals for either question. Students may need to adjust a setting described in most spreadsheet applications as a series to format the graph as requested in this problem.

d. Use the data collected from the class to evaluate a student's response. Students should develop their answers on the conditional frequencies of a right-handed person.

STUDENT PAGE 81

6. Collect the class data on eye dominance in a two-way table similar to the one below. In your table, fill in the values for cells a, b, c, and d.

	Eye Dominance	
	Left	**Right**
Hand Dominance **Left**	a	b
Right	c	d

a. Among the right-handed students, what is the conditional relative frequency that a person selected at random is right-eyed?

b. If a randomly selected student from your class is right-handed, what is the probability that this person has left-eye dominance?

c. Use a grid like the one below to make a bar graph showing the distribution of the conditional relative frequencies.

Right-Eyed Left-Eyed Right-Eyed Left-Eyed
 Right-Handed   Left-Handed

d. If a person is right-handed, do you think he or she is more likely to be left-eyed or right-eyed? Explain your answer.

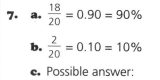

7. a. $\dfrac{18}{20} = 0.90 = 90\%$

b. $\dfrac{2}{20} = 0.10 = 10\%$

c. Possible answer:

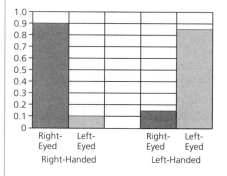

d. You would think that the person is more likely to be right-eyed. The conditional relative frequency of right-eyed people from the right-handed population is

$\dfrac{18}{20} = 0.90 = 90\%$. This indicates

that 90% of the right-handed population are right-eyed.

8. a. $\dfrac{11}{20} = 0.55 = 55\%$

b. $\dfrac{9}{20} = 0.45 = 45\%$

c. Possible answer:

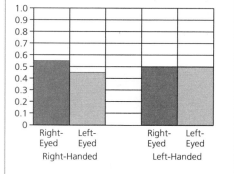

d. The data indicate that half of the left-handed people are left-eyed and half are right-eyed. As a result, estimating eye dominance for a left-handed person is a "50–50" estimate.

STUDENT PAGE 82

7. The two-way table below shows a different distribution of hand and eye dominance.

		Eye Dominance		
		Left	Right	Totals
Hand Dominance	Left	7	1	8
	Right	2	18	20
	Totals	9	19	28

a. Among the right-handed students in this distribution, what is the conditional relative frequency that this person is right-eyed?

b. Among the right-handed students in this distribution, what is the conditional relative frequency that this person is left-eyed?

c. Use a grid like the one in Problem 6 to make a bar graph showing the distribution of the conditional relative frequencies.

d. If a person from this distribution is right-handed, do you think he or she is more likely to be left-eyed or right-eyed? Explain your answer.

8. The two-way table below shows another distribution of hand and eye dominance.

		Eye Dominance		
		Left	Right	Totals
Hand Dominance	Left	4	4	8
	Right	9	11	20
	Totals	13	15	28

a. Among the right-handed students in this distribution, what is the conditional relative frequency that this person is right-eyed?

b. Among the right-handed students in this distribution, what is the conditional relative frequency that this person is left-eyed?

c. Use a grid like the one in Problem 6 to make a bar graph showing the distribution of the conditional relative frequencies.

d. If a person from this distribution is left-handed, do you think he or she is more likely to be left-eyed or right-eyed? Explain your answer.

STUDENT PAGE 83

9. Answers will depend on the data collected from the class. If a student indicates that the distribution is closer to that in Problem 7, then he or she is indicating that for left-handed people there is a greater percent with left-eyed dominance; similarly, for right-handed people there is a greater percent with right-eyed dominance. If a student indicates that the distribution is closer to that in Problem 8, then he or she is indicating that hand dominance does not relate to eye dominance. If a student indicates neither, either he or she does not understand the two examples, or else the class demonstrates other associations; that is, right-handed people are mostly left-eyed dominant and left-handed people are mostly right-eyed dominant.

10. **a.** The table in Problem 7 shows a strong association. The table in Problem 8 shows a weak or no association.

   b. Answers will depend on the specific data collected.

Problems 7 and 8 involve different distributions. The cells were set in Problem 7 to show that if a person is right-handed, then he or she is almost certain to be right-eyed. The cells in Problem 8 were set to show that if a person is right-handed, then there is no tendency toward being left-eyed or right-eyed.

9. Does it appear that your class distribution from Problem 6 is closer to the distribution shown in Problem 7, Problem 8, or neither problem? Explain your answer.

A relationship, or an association, exists if knowing the response to one of the variables helps to predict what the response might be to the other variable.

10. Study the tables in Problem 7 and Problem 8.
   a. Which table shows strong association? Which table shows weak or no association?
   b. Use your class distribution to determine whether there appears to be a strong relationship or association between hand dominance and eye dominance. Explain.

Summary

An association exists if knowing the response to one of the variables helps to predict what the response might be to the other variable. Knowing that person was right-handed helped predict his or her eye dominance. Knowing whether or not a person would walk out of a movie because of its violent content helped predict whether the person was a male or a female. If conditional relative frequencies for two groups are considerably different, then an association may exist.

Practice and Applications

11. a. $\frac{637}{1144} \approx 0.557 = 55.7\%$

b. $\frac{620}{1144} \approx 0.542 = 54.2\%$

c. $\frac{255}{524} \approx 0.487 = 48.7\%$

d. $\frac{382}{620} \approx 0.616 = 61.6\%$

e. Possible answer:

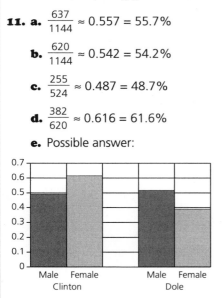

f. The bars are not proportional, indicating an association of the distribution. Further examination of the relative frequencies indicates that a higher percent of female voters supported Clinton.

12. The focus of the research was to determine attitudes along with practices. The project wanted to determine if students' perceptions of their peers were accurate. Studies attempting to evaluate perceptions against actual practices are important in addressing the problems examined through the survey.

STUDENT PAGE 84

Practice and Applications

11. On election day, exit polls are conducted by the media to determine who voted and why voters voted for a particular candidate. One such poll was conducted in 1996 by Voter News Service in Wisconsin. The results below are based on individual voter questionnaires after voters left polling places.

	Voted for		
	Clinton	Dole	Totals
Men	255	269	524
Women	382	238	620
Totals	637	507	1144

a. What percent of the voters voted for Clinton?

b. What percent of the voters were female?

c. Of the male voters, what percent voted for Clinton?

d. Of the female voters, what percent voted for Clinton?

e. Use a grid like that in Problem 6 to make a bar graph showing the distribution of conditional relative frequencies.

f. Use the conditional relative frequencies and the bar graph to comment on whether there seems to be an association between the sex of the Wisconsin voter and the candidate for whom the person voted.

12. The national Teenage Attitudes and Practices Survey obtained completed questionnaires from around the country by phone and mail from a randomly selected group of people 12–18 years old. The sample involved 9965 teenagers. Many of the questions asked the teenagers about their perception of the behavior of their peers. Why do you think the questions asked the students about their perceptions of peer behavior rather than ask questions about their own behavior?

Some of the questions on the Teenage Attitudes and Practices Survey asked the teenagers about their own behavior rather than their perceptions of their peers. The data shown are the results collected from the question "Do you believe cigarette smoking helps reduce stress?"

	Never Smoked	Experimented with Smoking	Former Smoker	Current Smoker
Yes	12.0%	18.7%	29.8%	46.5%
No	84.9%	78.5%	68.9%	51.7%
Don't Know	3.0%	2.5%	1.6%	1.6%

STUDENT PAGE 85

13. a. The percents were calculated using the totals from each column. In other words, the percent of any particular cell was based on the number in the cell and the total from the column involving the cell. If students indicate a concern that the percents do not add to 100%, indicate this is partially a result of rounding and the fact that some students within each category did not answer the smoking question.

b. 12.0% means 12% of the students who indicated they never smoked answered "yes" to the question "Do you believe cigarette smoking helps reduce stress?"

c. 12.0%

d. 46.5%

e. One would think his response would be "no," since 84.9% of those who never smoked said "no."

f. The paragraph can be written using the table in several ways. The table shows, however, an association. Students who have been involved in smoking or are current smokers indicate "yes" to the question in greater percents. This association should be discussed by the students in their paragraphs.

13. Use the data in the table on page 84 to answer the following questions.

   a. How were these percents calculated?

   b. Explain what the 12.0% represents.

   c. Among students who have never smoked, what percent believe that cigarette smoking helps reduce stress?

   d. Among students who are current smokers, what percent believe that cigarette smoking helps reduce stress?

   e. If you knew that a teenager never smoked, what do you think his response to the question "Do you believe cigarette smoking helps reduce stress?" would be?

   f. Are opinions on whether or not cigarette smoking helps reduce stress associated with the smoking status of the person responding? Write a short paragraph justifying your answer. Your paragraph should include appropriate data, calculations, and graphs.

Constructing Tables from Conditional Probabilities

Materials: Lessons 9 and 10 Quiz (optional)
Technology: graphing calculator, spreadsheet program (optional)
Pacing: 1 class period

Overview

In this lesson, students are asked to construct two-way tables. In the previous lessons, students were given the data in a table or calculated data from an experiment and placed the results in a table. This lesson presents results from a *Life* magazine survey and information concerning a test used to screen blood samples. Students are asked to use these examples to construct a two-way table and then calculate expected frequencies. Problems 1 and 2 use results from the *Life* magazine poll, and students begin placing data in a two-way table. Problems 3 and 4 present percents concerning the accuracy of a blood-screening test. Again, students are asked to convert the percents to expected values and the expected values are placed in a two-way table. In the remaining problems, data are presented in many different forms, but students are asked to complete a two-way table of expected values.

Teaching Notes

In Problems 1 and 2, as students complete the table, it is important that they realize that we would expect an equal number of men and women. But the expected marginal totals for the *Yes* and *No* columns come from the survey results. Students may need to be reminded that in Lesson 5 we used survey results to answer questions about the population. Students may

need some help with Problems 3 and 4. The term *false positive* can be confusing. After the table is constructed, point out which cells contain information that shows an error in testing. Problem 6 illustrates that not all data can be placed in a table. Students need to be aware that the data presented are relative frequencies; and unless they make an assumption of how many boys and girls took part in the survey, they cannot complete the table.

Follow-Up

Find other examples like the one in Problem 6 and discuss when the data can be organized in a two-way table and when they cannot be placed in a table. The *USA Today* newspaper provides many examples that can be used for this Follow-Up.

STUDENT PAGE 86

Solution Key

Discussion and Practice

1. a. Yes; the results are conditional. The phrase "two-thirds of the women interviewed" indicates that the frequencies were based on responses from the population of women. Similarly, "half of the men" indicates that the summary was based on responses from the men interviewed.

LESSON 10

Constructing Tables from Conditional Probabilities

How many hours of TV do you watch each day?

Do your friends watch TV more or fewer hours each day than you do?

OBJECTIVES

Determine expected frequencies from conditional probabilities.

Interpret probability statements by constructing an appropriate table of expected frequencies.

Determine unconditional probabilities from a table of expected frequencies.

INVESTIGATE

Information often comes to you as conditional relative frequencies or conditional probabilities, even though it is not usually labeled as such. A recent poll in *Time* magazine states that 20% of U.S. students watch more than 5 hours of TV per day, but the figure is only 14% for France and 5% for Canada. Is this conditional information? Yes, it is conditional on the countries involved. There are about 61 million students in U.S. schools, and so this conditional percent can be translated in the fact that about $61(0.20) = 12.2$ million students watch more than 5 hours of TV per day in this country. What would you need to know in order to find out how many students in France watch more than 5 hours of TV per day? Do you think it could be greater than the number for the United States?

Discussion and Practice

Constructing a Table from Poll Results

1. A poll in *Life* magazine entitled "If Women Ran America" reports that two thirds of the women interviewed say that the problem of unequal pay for equal work is a serious one, while only half of the men interviewed have this opinion.

a. Are the given poll results conditional information? Explain your answer.

STUDENT PAGE 87

b. Students would probably estimate that 1000 or one-half of the population are women. This is based on the assumption the distribution of men and women for a large population is approximately equal.

c. You would expect $\frac{2}{3}$ of the 1000 women to indicate this concern, or approximately 667 women.

d. You would expect $\frac{1}{2}$ of the men to indicate this concern. If 1000 of the people interviewed are men, then 500 of them would indicate this concern.

e.

Is Unequal Pay for Equal Work a Serious Problem?

	Yes	No	Totals
Men	500	500	1000
Women	667	333	1000
Totals	1167	833	2000

f.

Is Unequal Pay for Equal Work a Serious Problem?

	Yes	No	Totals
Men	500	500	$\frac{1000}{2000} = 0.50$
Women	667	333	$\frac{1000}{2000} = 0.50$
Totals	1167	833	$\frac{2000}{2000} = 1.00$

The marginal relative frequencies indicate that a student assumes an equal number of men and women in the sample.

b. About how many women would you expect to be in a group of 2000 Americans? Explain.

c. For a typical group of 2000 Americans, how many would you expect to be women who think that unequal pay for equal work is a serious problem?

d. For a typical group of 2000 Americans, how many would you expect to be men who think that unequal pay for equal work is a serious problem?

e. Copy the table below. Then use your results from parts b, c, and d to complete your table.

Is Unequal Pay for Equal Work a Serious Problem?

	Yes	No	Totals
Men			
Women			
Totals			2000

f. Find the marginal relative frequencies for the men and women rows and explain what each frequency represents.

2. In the same poll, one half of the women and one third of the men thought that discrimination in promotions was a very serious problem.

a. Copy and complete the following table to show how a typical group of 1000 Americans would be divided on this issue. Remember, the total numbers of men and women are approximately equal.

Is Discrimination in Promotions a Serious Problem?

	Yes	No	Totals
Men			
Women			
Totals			1000

b. Approximately what fraction of Americans think that discrimination in promotions is a very serious problem?

2. a.

Is Discrimination in Promotions a Serious Problem?

	Yes	No	Totals
Men	167	333	500
Women	250	250	500
Totals	417	583	1000

b. With the estimated values from the previous table, the fraction of Americans who think discrimination in promotions is serious is $\frac{417}{1000} = 0.417$, or 41.7%.

STUDENT PAGE 88

3. The statement is conditional, as the percent is based on the blood samples that contain HIV.

4. **a.** 500 samples contain HIV. ELISA indicates 99% of them as positive; that is, $0.99 \times 500 = 495$ samples are positive.

b. 500 samples contain HIV. ELISA indicates 2% of them as positive for a false positive reading; that is, $0.02 \times 500 = 10$ samples are false positive.

c.

	Contain HIV	No HIV	Totals
Tested HIV Positive	495	10	505
Tested HIV Negative	5	490	495
Totals	500	500	1000

d. The false positives are indicated in the cell with 10 samples.

e. The other cell indicating an error would be the 5 samples contain HIV that are tested as HIV negative.

f. The predictive value is $\frac{495}{505} \approx 0.980 = 98\%$.

This percent is close to 100%, indicating a very good screening of the blood samples.

Blood Tests and Conditional Probability

ELISA is a popular test for screening blood samples for the presence of HIV. For blood samples known to contain HIV, ELISA shows a positive result 99% of the time. This means that 99% of the time the test will correctly identify blood that contains HIV. For blood samples known to be free of HIV, ELISA still reports a positive result about 2% of the time. This means that 2% of the time the test will incorrectly report that a blood sample contains HIV. This is called a *false positive*.

3. Is the 99% stated above conditional or unconditional information? Explain.

4. Assume that a medical lab tests 1000 blood samples using ELISA. Also assume that 50% of the samples contain HIV.

  a. How many blood samples known to contain HIV would you expect to test positive?

  b. How many false positives would you expect to see?

  c. Copy and complete the following table by filling in the remaining expected frequencies.

	Contain HIV	No HIV	Totals
Tested HIV Positive			
Tested HIV Negative			
Totals	500	500	1000

  d. Which cell contains the false positives?

  e. Which other cells show tests that are in error?

  f. One measure of the accuracy of the screening procedure is to examine the probability that an HIV sample will test positive. This is called the **predictive value** of the procedure. This means that the predictive value from the data in the table above would be found by dividing the number of samples that test positive and contain HIV by the total number of positive tests. Find this value. Would you say the test is doing a good job of screening? Why or why not?

STUDENT PAGE 89

5. **a.** The results are conditional, as each is based on a subtotal of the population surveyed.

b. You would expect half of the entire group of children to be girls. Therefore, you would estimate that 400 children are girls.

c. 44% of this group collect cards: $0.44 \times 800 = 352$ children

d. 25% of the children who collect cards are girls: $0.25 \times 352 = 88$ girls

e.

Do You Collect Sports Trading Cards?

	Yes	No	Totals
Boys	264	136	400
Girls	88	312	400
Totals	352	448	800

Summary

Information is often presented in terms of conditional relative frequencies, which can be interpreted as conditional probabilities. To understand these conditional statements and to obtain unconditional information from them, it is helpful to think in terms of expected frequencies for a typical sample from the population under investigation. These expected frequencies are most easily understood when displayed in a two-way table.

Practice and Applications

5. A *Sports Illustrated for Kids* Omnibus Survey reported that about 44% of children aged 9–13 collect sports trading cards. Of this group who collected cards, about 1 in 4 of them are girls.

a. Are the given survey results conditional information? Explain.

b. About how many girls aged 9–13 would you expect to be in a random group of 800 children aged 9–13? Explain.

c. Of the 800 children surveyed, how many said that they collect sports trading cards?

d. Of the 800 children surveyed, how many girls said that they collect trading cards?

e. Copy the table below. Then write the expected frequencies in the appropriate cells of your table.

Do You Collect Sports Trading Cards?

	Yes	No	Totals
Boys			
Girls			
Totals			800

STUDENT PAGE 90

6. a. The poll results are conditional. Each percent is based on either the population of girls or the population of boys.

b. 49.2% of the girls aged 9–13 play video games less than 1 hour per day.

c. It is possible to complete a *conditional relative frequency chart* given the data. It is not possible to complete a frequency chart without knowing the total number of students polled. Table A below represents the conditional relative frequencies based on the graph.

If you assumed that of the children polled an equal number were boys and girls (a reasonable assumption), then you could complete the total percents for each column. One way to do this is to consider a sample of 200 children and assume 100 are girls and 100 are boys. Table B below gives the corresponding frequencies. Marginal relative frequencies are given in Table C below.

6. The results of a *Sports Illustrated For Kids* poll, reported in *USA Today*, October 4, 1995, are shown in the graph below.

Number of Hours per Day Boys and Girls Aged 9–13 Play Video Games

a. Are the given poll results conditional information? Explain.

b. What does the 49.2% represent?

c. A student wanted to organize the results of the poll in a two-way table similar to the one that follows. Is it possible to complete the table from the data given? Can you fill in any of the marginal or joint relative frequencies? If so, which ones?

Time Playing Video Games

	2 or More Hours	1 Hour	Less than 1 Hour	Totals
Boys				
Girls				
Totals				

7. In all states, statewide exit polls are conducted on election day to help the news media predict which candidate will win the election. The results shown below are based on 1300 voters in Wisconsin on Election Day, 1996.

Percent of Total		Voted for			
		Clinton	Dole	Perot	Other
47%	Men	41%	44%	11%	4%
53%	Women	55%	34%	8%	3%
36%	Democrat	86%	6%	6%	2%
34%	Republican	13%	80%	6%	1%
30%	Independent/Other	42%	31%	20%	7%

A. **Time Playing Video Games (hours)**

	2 or More	1	Less than 1	Totals
Boys	36.9%	33.6%	29.5%	100%
Girls	20.1%	30.8%	49.2%	100%*
Totals				

* Total percent is slightly more than 100% due to rounding.

B. **Time Playing Video Games (hours)**

	2 or More	1	Less than 1	Totals
Boys	37	34	29	100
Girls	20	31	49	100
Totals	57	65	78	200

C. **Time Playing Video Games (hours)**

	2 or More	1	Less than 1	Totals
Boys	36.9%	33.6%	29.5%	100%
Girls	20.1%	30.8%	49.2%	100%
Totals	28.5%	32.5%	39%	100%

STUDENT PAGE 91

7. a. Results may differ slightly because of rounding. See table below.

b. $\dfrac{623}{1300} \approx 0.479 = 47.9\%$

c. $\dfrac{402}{468} \approx 0.860 = 86.0\%$

d. $\dfrac{402}{623} \approx 0.645 = 64.5\%$

e. See table below.

f. $\dfrac{379}{689} \approx 0.550 = 55.0\%$

g. $\dfrac{379}{630} \approx 0.602 = 60.2\%$

a. Copy the following table. Then use the data on page 90 to complete your table.

| | Voted for | | | | |
	Clinton	Dole	Perot	Other	Totals
Democrat					
Republican					
Independent/Other					
Totals					1300

b. If a Wisconsin voter is chosen at random, what is the probability that the person voted for Clinton?

c. If a Wisconsin voter is chosen at random from among those who said they are Democrats, what is the probability that the person voted for Clinton?

d. If a Wisconsin voter is chosen at random from among those who voted for Clinton, what is the probability that the person is a Democrat?

e. Use the data from part a to complete, as far as possible, the following table of frequencies of voters calculated from the exit-poll results.

| | Voted for | | | | |
	Clinton	Dole	Perot	Other	Totals
Men					
Women					
Totals					1300

f. If a Wisconsin voter is chosen at random from among the women voters, what is the probability that the woman voted for Clinton?

g. If a Wisconsin voter is chosen at random from among those who voted for Clinton, what is the probability that the person is a woman?

a.

	Clinton	Dole	Perot	Other	Totals
Democrats	402	28	28	10	468
Republicans	57	354	27	4	442
Independents/Other	164	121	78	27	390
Totals	623	503	133	41	1300

e.

	Clinton	Dole	Perot	Other	Totals
Men	251	269	67	24	611
Women	379	234	55	21	689
Totals	630	503	122	45	1300

LESSON 11

Comparing Observed and Expected Values

Materials: none
Technology: graphing calculator
Pacing: 2 class periods

Overview

In Lesson 9, students used their observations of conditional relative frequencies and a bar graph of these frequencies to make a decision regarding association. In this lesson, students compare the results of a simulation to results from a survey to help them decide if two variables are independent. In the first part of the lesson, students design a simulation to randomly pick a 1, a 2, or a 3 for 72 trials. A success is determined if a 1 is picked. The number of successes out of 72 trials is tallied and class data are collected. The main idea is to see how often the number of successes that occurred by chance compares with the 42 (58% of 72) successes observed in the study. If the number observed in the study occurs fairly often by chance, then we would conclude that there is nothing special that occurred in the study.

Problems 4–7 use results from a simulation to decide if two variables are independent. The simulation is based on the assumption that the two questions are unrelated or independent. Under this assumption, the expected value of one cell is calculated. Then a simulation is designed to determine how likely it is that the observed value occurs by chance.

The remaining problems provide two more opportunities for students to design and carry out a simulation to help decide if two variables are independent.

Teaching Notes

Students find the first simulation relatively easy to set up and carry out. The main emphasis is that we are comparing what would happen by chance (the results of the simulation) to what was observed. If the observed value occurs by chance fairly often, then we conclude that the observed value is *not* significant. The general rule of thumb is that if the observed value occurs by chance less than 5% of the time, it is considered significant.

The second simulation is much more difficult for students to understand. Instead of using 50 slips of paper, you could give each group of students a deck of cards. Students would designate 20 red cards as representing the 20 "yes" responses to the health question and 30 black cards as the 30 "no" responses to the health question. The cards are shuffled and dealt into two piles, one pile of 36 representing the 36 "yes" responses to the breakfast question and another pile of 14 representing the 14 "no" responses to the breakfast question. Students count the number of red cards in the pile of 36 and record this value in the Yes-Yes cell. You may want to demonstrate to students that once they know the value in this cell, they need not consider what is in the other pile of cards.

Follow-Up

You may want to use the data collected on the hand and eye dominance and have students design and carry out a simulation to decide if there is an association between the two variables. The simulation would be run under the assumption of no association.

STUDENT PAGE 92

Solution Key

Discussion and Practice

1. **a.** $\frac{1}{3} \approx 0.333 = 33.3\%$

   b. Answers will vary; however, it is anticipated that students will comment that 58% is considerably higher than the 33.3% determined by chance.

LESSON 11

Comparing Observed and Expected Values

If you tossed a coin 100 times and you counted 35 heads, would you be surprised by the results?

If you took a sample of 500 people living in your community and 280 men and 220 women were chosen for your survey, would you think that your survey was not random?

OBJECTIVES

Simulate the variability that may be attached to a table of expected values.

Make decisions about the presence of association based on this variability.

In this lesson, you will compare data collected by a survey or an experiment to what you would have expected if the results happened purely by chance.

INVESTIGATE

Love Is Not Blind

"Love is not blind, and study finds it touching" is the headline of an article in the *Gainesville Sun,* June 22, 1992. Seventy-two blindfolded people each tried to distinguish his or her partner from a group of three people, one who was actually the partner, by touching the forehead. The blindfolded people making the selections were correct 58% of the time.

Discussion and Practice

1. Consider the data in the paragraph above.

   a. If a person picked his or her partner by chance, what percent of the time would you expect the person to be correct?

   b. Does the observed 58% success rate seem to be far from what you would expect due to chance?

STUDENT PAGE 93

2. Answers will vary. Students might comment that none of the participants made a correct response. Similarly, they might comment that all 72 made the correct response. Therefore, intervals covering 0 to 72 are possible. It is very unlikely the extremely low and the extremely high values will occur; therefore, students may indicate that smaller ranges, for example, 15 to 35, are more reasonable.

Many statistical problems have the same structure as this experiment. Data are observed under certain experimental conditions. The question is "What is the probability that the observed data could have happened by chance?" To answer this question, you will design and run a simulation. The simulation model provides probabilities based on the occurrences that happen by chance. After the simulation is run, the observed data are compared to the results produced by the simulation. If they agree, then you could conclude that the observed data might have happened by chance. If the observed data and the simulated results do not agree, then you might conclude that the observed data represent a significant result.

Assume that the blindfolded participants in the study are merely guessing about their partners. With three possible choices for each blindfolded participant, the chance of choosing the correct partner by guessing is one third, or about 33%. How does the observed 58% compare with the 33% by chance? Is it a significant result?

It will be easier to work with frequencies rather than with percents. The number of correct decisions observed was 58% of 72, or 42. The number expected due to chance would be $\frac{1}{3}$ of 72, or 24. However, even if the participants were merely guessing, they might have had more or fewer than exactly 24 correct decisions.

2. If all 72 participants were guessing, what do you think is a reasonable interval of values that might contain the number of correct decisions?

One way to answer Problem 2 is to design and run a *simulation.* Simulation involves using a random number to model the behavior of the real event under investigation. The real event of choosing the correct partner purely by chance has a probability of $\frac{1}{3}$. To model this probability, random numbers 1, 2, and 3 can be used, designating 1 to represent "success" (choosing the right partner) and 2 and 3 to represent "failure" (*not* choosing the right partner).

3. a. You may want to review with students a few calculator instructions to create a random distribution. For example, if students use a TI-83, they should select **MATH**. From this menu, they should select **PRB** and **5:randInt(**. Then they should hit **ENTER** to paste this instruction in the main window and complete the following: **randInt(1,3)**

Repeated selections of **ENTER** will generate random selections of a 1, 2, or 3. To generate 72 random selections with one command, enter **randInt(1,3,72)**. These selections could be used for a student to carry out the simulation. For some students, developing a program to carry out this instruction 72 times while counting the times a 1 occurs is a reasonable challenge. It will also generate numerous results for the next part of this problem.

Answers will vary based on the outcomes to the simulation. Expect the total number of successes to be close to 24.

b. Possible answer:

c. The shape of the graph is erratic; however, the outcomes peak at 24 and then taper off. The expected value 24 is marked with a vertical line. This outcome does appear to be a center of the simulation. For the above example, 19 outcomes were less than 24 and 11 outcomes were greater than 24. Answers will vary according to the data collected by the students.

STUDENT PAGE 94

3. Use a simulation to answer Problem 2.

a. Randomly select 1, 2, or 3 and record the outcome as "S" for success (1—picked the correct partner) or "F" for failure (2 or 3—did not pick the correct partner). Since the real experiment contacted 72 people picking their partners, you will need to choose a random number 72 times. Work with your group to randomly select a 1, 2, or 3 a total of 72 times. When finished, count the successes. Was your value close to 24, the expected number?

b. Repeat the 72 selections of random numbers 4 more times so you have 5 values for the number of success out of 72 decisions. Combine these 5 values with those from other groups in your class. Construct a graph of the class data. Use a number line like the one below.

15 16 17 18 19 20 21 22 23 24 25 26 27 28 29 30 31 32 33 34 35

c. Describe the shape of the graph you developed in part b. Mark the expected number of 24 on the number line. Do the points center around this value?

d. Compare the data from the experiment to the 42 successes reported in the research. Did anyone in class achieve 42 successes? What fraction of the data points on the graph are greater than or equal to 42?

e. Do you think people can choose their partners by touching them on the forehead; or do you think the results of the experiment happened by chance? Explain.

Independence Between Two Variables

4. The question of healthy diet versus eating breakfast was introduced in Lesson 5. The two-way table below shows the marginal totals.

| | **Do You Think Your Diet Is Healthy?** | | |
	Yes	**No**	**Totals**
Do You Eat Breakfast at Least 3 Times a Week? **Yes**			36
No			14
Totals	20	30	50

---

d. 42 successes did not occur in any of the examples. The graph indicates that the value 31, the greatest number of successes, occurred once. Comparing this outcome to 42 indicates that it is very unlikely that 42 would occur by chance.

e. Expect students to answer "yes," as the outcome of 42 is very rare in the simulations conducted by students.

STUDENT PAGE 95

4. a. 36 out of 50, or 72%

b. 20 out of 50, or 40%

c. 40% of 1200, or 480 students

5. a. See table below.

b. This value, 8.4, represents the number of students who answered "no" to the diet question and "no" to the breakfast question if the two questions are unrelated (independent). It represents approximately 28% of the 30 students who answered "no" to the diet question.

6. See table below.

a. What fraction of the 50 polled students eat breakfast at least 3 times a week?

b. What fraction of the polled students think their diet is healthy?

c. If there are 1200 students in the school for which the poll was conducted, how many would you expect to say that they have a healthy diet?

5. Suppose that the two questions are not related. The statistical term we use to describe this is ***independence.*** If the assumption that the questions are not related or independent for the students surveyed, then answering "yes" to the breakfast question will not affect the response to the diet question. In other words, the percent answering "yes" to the diet question should be about the same for each group (each row of the table) in the breakfast question. Since $\frac{20}{50}$, or 0.40, answered "yes" to the diet question, then among the students who answered "yes" to the breakfast question, you should expect 40% of 36, or 14.4, to have answered "yes" to the diet question.

a. In a table like the one above, place 14.4 in the upper left-hand corner of the table (the "yes-yes" cell). Complete your table and describe your strategy.

b. Describe what the value in the lower right-hand corner of your table (the "no-no" cell) represents.

6. The table below shows the original data table as presented in Lesson 5. Copy and complete the rest of the table using only the information presented.

	Do You Think Your Diet Is Healthy?		
	Yes	**No**	**Totals**
Yes	17		36
No			
Totals	20		50

(rows labeled: **Do You Eat Breakfast at Least 3 Times a Week?**)

The goal now is to see if the two questions are independent, or not related, by comparing the observed results (the table of original data) to the table of expected values. Since the joint information inside the table can be determined by knowing the marginal totals and one cell, the problem of determining whether or not the two variables are independent can focus on one cell. The problem can be restated as follows:

5. a.

Do You Think Your Diet Is Healthy?

		Yes	No	Totals
Do You Eat Breakfast at Least 3 Times a Week?	**Yes**	14.4	21.6	36
	No	5.6	8.4	14
	Totals	20	30	50

6.

Do You Think Your Diet Is Healthy?

		Yes	No	Totals
Do You Eat Breakfast at Least 3 Times a Week?	**Yes**	17	19	36
	No	3	11	14
	Totals	20	30	50

STUDENT PAGE 96

7. This is an excellent simulation to develop with students. The topic of an independent versus dependent distribution is not readily understood by students. The steps presented should provide a workable set of data to explain this topic.

a. Answers depend on the data set collected. 36 values collected by a group of students are represented in the following line graph.

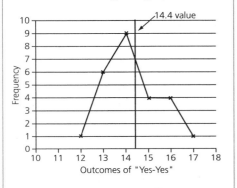

b. The shape peaks around the expected value of 14 to 15. A vertical line marking the expected value of 14.4 is drawn on the line graph.

Is the 17 so far from the expected frequency of 14.4 for that cell that its occurrence probably did not happen by pure chance?

7. The chance that a frequency of 17 or more could occur in that cell under the assumption of independence can be approximated by a simulation. The following steps lead through a simulation to help find this probability.

Simulation Steps

i. Use 50 slips of paper of equal size. Mark 20 of the slips "Y" to represent those that answered "yes" to the diet question and mark 30 "N" for those that said "no" to the diet question.

ii. Mix the 50 slips of paper in a box and randomly select 36 of the slips. These represent the 36 students who said "yes" to the breakfast question. Under the assumption that the questions are independent, these 36 should behave like a random sample from the whole group of 50. That is, the proportion of the 36 who say "yes" should be about the same as the proportion of the 50 who said "yes."

iii. Count the slips among the sample of 36 that have a "Y." This represents the number of students who also said "yes" to the diet question from those who said "yes" to the breakfast question. In other words, this represents the number who said "yes" to both questions.

iv. Repeat this procedure 5 to 10 times to obtain a set of possible values for the "yes-yes" cell.

a. Combine your data with the data collected by other groups in your class. Use the number line like the one below to show the distribution of frequencies for the "yes-yes" cell.

7 8 9 10 11 12 13 14 15 16 17 18 19 20 21 22

b. Describe the shape of the graph produced. Make the expected value of 14.4 on your graph. Do the points center around this value?

STUDENT PAGE 97

(7) c. Using the outcomes of the simulation presented in the line graph, the outcome of 17 occurred only twice from the 36 trials. Although the number of times will vary depending on the data collected from the class, it is anticipated that 17 will not occur often. The estimate of the probability of seeing 17 or more would be very small.

d. As the probability of a 17 or more is very small, it is not likely this result occurred by chance. Outcomes within the range of 13 to 16 would be the most likely to occur by chance.

e. Given the results from the simulation, students would conclude an association exists between the two questions. This associations indicates a connection or relationship between the questions.

c. Compare the observed value of 17 to the class data from the simulation. Does the value of 17 occur very often? What is your estimate of the probability of seeing 17 or more in the "yes-yes" cell?

d. Does it seem reasonable to say that the result of 17 or more is quite likely to occur by chance? Explain.

If you answered "yes" to part d, then you are suggesting that the two questions are independent; that is, the independent model agrees with the data, and there is probably no association. Generally, if the probability that an event will occur is less than 10%, the investigator may decide that this is a rare event and, therefore, there may be an association between the two variables in question.

e. What do you conclude about the association between questions dealing with diet and breakfast at Rufus King High School?

Summary

One of the goals of a statistical investigation is to compare data with theoretical models to see if the data support the model. For example, to check a coin for balance, one could assume that the coin is balanced (the model), toss it many times to observe the fraction of heads (the data), and compare the observed fraction of heads to $\frac{1}{2}$, the ratio assumed for this model. If the observed fraction of heads differs considerably from $\frac{1}{2}$, then the model of a balanced coin is questioned.

To decide if two questions or variables are associated, you should construct a two-way table of expected frequencies under the assumption that the two variables are independent. Design and run a simulation based on a model of independence. Compare the results of the simulation with the observed data and find an estimate for the probability that the observed data could have occurred by chance. If this probability is small, usually less than 10%, you may conclude that there is an association between the two variables.

Simulating the distribution of results that could be obtained from the model allows the investigator to see if the original outcome is likely or unlikely under the assumptions of the model. A simulated distribution of possible outcomes under the model is a powerful tool for decision making.

STUDENT PAGE 98

8. a. See the table below for observed frequencies.

b. Using the marginal totals as described to the students, the following proportion is set up and used to complete the table of expected frequencies:

$$\frac{\text{expected number of "yes-yes"}}{\text{number of students who own a vehicle}}$$

$$= \frac{\text{number of students with outside job}}{\text{total number of students}}$$

Let x = the expected number of the "yes-yes" cell. Using x with the specific marginal totals yields the following proportion: $\frac{x}{20} = \frac{26}{40}$

The value of x is 13, which results in the table for expected frequencies below.

Practice and Applications

8. A sample of 40 students were asked two questions:

Do you have a job outside of school?

Do you have a car or truck of your own?

Half of the students had a car or truck of their own. Twenty-six of the 40 students had jobs. Twelve of the students did not have a job and did not own a vehicle.

a. Copy and complete the two-way table of observed frequencies.

Observed Frequency Table

		Job Outside of School		
		Yes	No	Totals
Do You Own a Car or Truck?	Yes			
	No			
	Totals			40

b. To decide if these two questions are associated, we need to construct a two-way table of expected frequencies. Assume that the two questions are independent of each other and copy and complete the following table. Remember that the marginal totals are the same as found in the observed frequency table. Also, since you are assuming independence, the ratio of the number in the "yes-yes" cell to the total of the number of students who own a car or truck should be the same as the ratio of the number of students who have a job to 40, the total number of students in the sample.

Expected Frequency Table

		Job Outside of School		
		Yes	No	Totals
Do You Own a Car or Truck?	Yes			
	No			
	Totals			40

c. Do the values in the observed-frequency table seem to be far from the corresponding values in the expected-frequency table?

d. Based on the tables constructed above, does it appear that the two questions are associated?

---

8. a.

Do You Own a Car or Truck?	Job Outside of School?		
	Yes	No	Totals
Yes	18	2	20
No	8	12	20
Totals	26	14	40

8. b.

Do You Own a Car or Truck?	Job Outside of School?		
	Yes	No	Totals
Yes	13	7	20
No	13	7	20
Totals	26	14	40

c. Based on the results from the simulation developed in Problem 7, it is anticipated the students will indicate the expected value of 13 and the observed value of 18 are quite different.

d. The greater value observed would suggest the questions are associated; therefore, the collected data suggests a connection between owning a car or truck and having a job outside of school.

STUDENT PAGE 99

9. a. Students should follow the steps outlined in the problem. Students will mark 26 slips with a "Y," representing the 26 students who have a job outside of school. The remaining slips will be marked with an "N." After thoroughly mixing up the slips, they will proceed to pick 20 slips representing the 20 students who own a car or truck, or answered "yes" to the second question. They will then record the number of slips marked "Y" from this selected group of 20. Students should repeat this process several times and combine their results. The combined collection should provide a large enough representation to estimate the association suggested by the two tables.

b. The simulation should indicate outcomes build up to values around 13. Outcomes slightly less or slightly more will occur and may represent the most frequent outcomes. The outcome of 18, however, is not anticipated to occur often, if at all. This suggests to students that the observed results appear to be unusual and that 18 would not occur by chance. An association between the two questions is suggested by this simulation. As the number of students who own a car or truck and have a job is greater than expected, students could conclude that having a job indicates a greater number of students own a car or truck, or owning a car or truck indicates that a greater number of students have a job.

10. a. The information reported produces the following observed frequencies.

---

9. To help answer Problem 8d, you need to find the probability that the observed value in the "yes-yes" cell could happen by chance under the assumption that the two questions are independent. To do this, we will design and run a simulation.

Simulation Steps

i. Prepare 40 slips of paper of equal size. Mark the appropriate number "Y." These should represent the number of students who answered "yes" they have a job. The rest of the slips should be marked "N."

ii. Randomly select the number of slips equal to the number of students who said that they own a car or truck. Record the number of "Y"s.

iii. Repeat this procedure several times. Combine your data with the data from the rest of your class. Construct a graph of the frequencies of "Y"s.

a. Determine the probability that the observed value in the "yes-yes" cell could happen by chance under the assumption that the two questions are independent.

b. Do the observed results in the original survey appear to be unusual under the independence assumption? What does that tell you about the association between the two questions? Does information on job status provide any information on whether or not a student might own a car or truck?

10. In March of 1995, a poll taken by the Gallup Organization for *U.S. News and World Report* reported that 43% of women and 34% percent of men said that they have been accused of being a "back-seat driver."

a. Assume that there were 50 men and 50 women in the survey. Copy and complete the following two-way table of observed frequencies.

Accused of Being a "Back-Seat Driver"

	Yes	No	Totals
Men			50
Women			50
Totals			100

Accused of Being a "Back-Seat Driver"

	Yes	No	Totals
Men	17	33	50
Women	21.5*	28.5*	50
Totals	38.5*	61.5*	100

*Note: Clearly the actual sample size for this survey was not 100 as the fractional values would not have occurred. It is appropriate to adjust the results so no fraction is involved. This adjustment requires rounding up for one cell and rounding down in the other as the marginal totals must be kept at the values indicated.

The information reported represents conditional relative frequencies. It might be necessary to remind students about their previous work with this topic.

STUDENT PAGE 100

b.

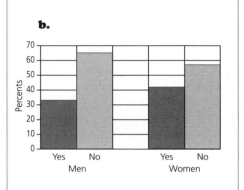

c. Answers will vary. Students are comparing 17 men to 21.5 (or 22) women who indicated they are accused of being a "back-seat driver." The difference between these two numbers is great enough to suggest that there may be an association between gender and the driving question. This difference, however, is not great enough to make a strong conclusion about the association.

d. If the variables are independent, then the proportion of women who indicated that they were accused of being a "back-seat driver" is the same as the proportion of men. This means the percent of men and women answering this question as "yes" would be the same.

e. The expected frequencies are represented in the following table. The "yes-yes" cell in this problem is the "men-yes" cell. Setting up the proportion given the marginal totals indicates that you would expect 38.5% of the 50 males to answer "yes" to this question. This value assumes that gender and the response to the driving question are independent.

b. Make a bar graph showing the distribution of the conditional relative frequencies. Uses a grid similar to the following.

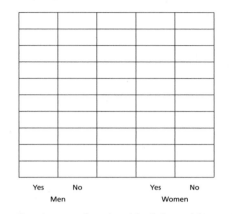

c. Does it appear from the table of observed frequencies that there is an association between gender and whether or not the person was accused of being a back-seat driver? Explain.

d. What does it mean for the two variables to be independent?

e. Copy and complete the following two-way table of expected frequencies. What assumptions did you use to make your calculations?

	Accused of Being a "Back-Seat Driver"		
	Yes	No	Totals
Men			50
Women			50
Totals			100

f. After completing the table of expected frequencies, would you change your answer to part c?

	Accused of Being a "Back-Seat Driver"		
	Yes	**No**	**Totals**
Men	19.25	30.75	50
Women	19.25	30.75	50
Totals	38.5	61.5	100

f. Answers will vary. The difference between 17 and 19.25 is not very great; therefore, students may indicate association or no association between gender and the driving question. Students are expected to justify their answers based on the chance of getting a 17 when the expected value is 19.25.

STUDENT PAGE 101

g. Students are expected to indicate steps similar to those in the previous problems. The fractions in the problems cause a new difficulty. To generate a procedure similar to the previous examples, students should let the 38.5 represent 38 and the 19.25 represent 19. The students can then create 100 slips of paper with 38 marked with a "Y" and 62 marked with an "N." After thoroughly mixing the slips, they should select 50 slips representing 50 males. The result for the trial would be the number of slips marked "Y." They should then record the result and repeat the process at least 25 times. It is suggested that students combine their results to increase the number of trials.

The probability of getting the value 17 should be answered on the basis of the number of times 17 occurred in the simulation. This result probably occurred enough times to suggest that the outcome of 17 was a result of chance. Adjust responses, however, based on the specific data recorded by the students.

h. The paragraph students develop should indicate the range of outcomes of the simulation and the probability of getting a 17 from the simulation. If this probability is under 10%, students might say that the gender and the driving question would indicate an association. If the probability is greater than 10%, students would probably indicate no association. In each case, students are indicating whether or not the outcome of 17 was likely or not likely to occur by chance.

g. Design and run a simulation to confirm your opinion of whether or not there is an association between the two variables. Write out the steps of your simulation and show the results of the simulation in graphical form. You should also state the probability of the observed frequency occurring by chance.

h. Write a paragraph explaining your conclusion on the association between gender and whether or not the person was accused of being a back-seat driver.

Assessment for Unit IV

Materials: deck of cards or slips of paper, End-of-Module Test
Technology: graphing calculator
Pacing: 1 class period or homework

This assessment enables you to evaluate students' understanding of Lessons 9–11. Problems 1 and 4 cover material from Lesson 10. Problems 2 and 3 cover material from Lesson 9. Problems 5 and 6 cover material from all three lessons.

Teaching Notes

As your students work through the problems, stress the need to clearly show their work and their thinking.

Problem 5 asks the students only to describe how they would design and run a simulation. The size of the numbers would make this a very time-consuming simulation. Problem 6 asks the students to design and run a simulation. You may wish to set the number of trials so that students know when they are finished with the simulation.

Follow-Up

After students have completed Problem 6, class results could be compiled and compared.

Solution Key

1. a. Yes. Each percent is based on a subgroup of the total population.

b. You would expect 50% of the 806 randomly selected adults would be women, or

$0.50 \times 806 = 403$ women

c. You would expect 43% of the women to be in this group, or

$0.43 \times 403 = 173.29 \approx 173$ women

d. The table below gives rounded calculations.

Accused of Being a "Back-Seat Driver"

	Yes	No	Totals
Men	173	230	403
Women	137	266	403
Totals	310	496	806

2. a. The percents are conditional. Although the information presented in this problem is similar to students' previous work with tables, this example develops the percents based on the totals of the columns. If students determine the sum of each column, they will notice that the values are close to, but not exactly, 100%. This is again due to rounding and the several responses that were not counted in the categories of "Yes," "No," or "Don't Know."

ASSESSMENT

Assessment for Unit IV

OBJECTIVE

Apply the concepts of conditional probability and simulation to decide if there is an association between two variables.

1. A poll, taken by the Gallup Organization for *U.S. News and World Report* in March, 1995, reported that 43% of women surveyed and 34% of men surveyed said that they have been accused of being a "back-seat driver."

a. Are the given poll results conditional information? Explain.

b. The poll consisted of 806 randomly selected adults. How many women would you expect to be in this group of 806?

c. How many women would you expect to have said that they have been accused of being a back-seat driver?

d. Complete the two-way table below.

Accused of Being a "Back-Seat Driver"

	Yes	No	Totals
Men			
Women			
Totals			

2. A question on the Teenage Attitudes and Practices Survey was "Do you believe almost all doctors are strongly against smoking?" Responses to the question are recorded in percents.

Do You Believe Almost All Doctors Are Strongly Against Smoking?

	Never Smoked	Experimented with Smoking	Former Smoker	Current Smoker
Yes	80.1%	78.8%	80.1%	80.5%
No	17.3%	18.8%	17.3%	16.7%
Don't Know	2.5%	2.3%	2.6%	2.6%

a. Are the percents joint, marginal, or conditional? Explain.

b. Without data describing actual frequencies, a specific value of the percent of teenagers who think doctors are strongly against smoking is not possible. An estimate, however, is possible, as the categories representing the teenage population all responded within the range of 78.8% to 80.5% to this question. This range could be used to indicate an estimate.

c. The fact that all categories describing teenagers—"Never Smoked," "Experimented with Smoking," and so on—have percents similar of doctors against smoking indicates the two items are independent or not associated. To indicate association, a greater difference would be required within the descriptions of teenagers and the "yes" response to the doctors' opinion.

3. **a.** Marginal percents are represented in the "Total" column; conditional percents are presented in the "Men" and "Women" columns. Students can determine the marginal values, as this column adds to 100%. The conditional percents are also indicated, in that the columns of "Men" and "Women" add to 100%.

b. The total population surveyed probably included a population equally represented by men and women. As a result, an average or mean of the percent of women and men for each of the categories representing the health question would represent the percent for the "Totals."

c. 58.1%; this is a challenging problem. If necessary, direct students to consider a population of 200 people, 100 women and 100 men. Given the data, students should be able to determine the frequencies in the following table.

STUDENT PAGE 103

b. From the table on page 102, can you approximate the percent of teenagers who think that almost all doctors are strongly against smoking? Explain your reasoning.

c. Are the opinions that doctors are against smoking associated with the smoking status of the person responding? Justify your answer with appropriate references to conditional relative frequencies. Show your work.

3. The following summary of a poll reported in the January 30, 1995, issue of *Time* magazine is given in percent form.

POLL RESULTS

How Much Effort Are You Making to Eat a Healthy and Nutritionally Balanced Diet?

	Men	Women	Totals
Very Serious Effort	31%	43%	37%
Somewhat Serious Effort	47%	43%	45%
Not Very Serious Effort	12%	9%	10%
Don't Really Try	10%	5%	8%

a. Are these joint, marginal, or conditional percents? How can you tell?

b. Why are the percents for totals midway between the percents for men and for women? Explain your reasoning.

c. From a group of people who say they are very serious about eating a balanced diet, one person is randomly selected. What is the approximate probability that this person will be female?

d. Is there a strong association between gender and efforts at eating a balanced diet? Justify your answer with appropriate calculations of conditional relative frequencies.

4. The Koop Foundation conducted a survey of 1600 urban residents. The survey, reported in the *Milwaukee Journal Sentinel* on December 2, 1995, found that 42% of Americans watch 3 or more hours of television a day. Of those watching this amount of television, 62% were from families with an income of $25,000 or less, compared with 38% from families earning more than $25,000.

a. Are the results given from the Koop Foundation survey conditional information? Explain your answer.

Effort at a Healthy Diet

	Very Serious Effort	Other Responses	Totals
Men	31	69	100
Women	43	57	100
Totals	74	126	200

The question asks for the conditional probability of selecting a female given that the person is very serious about efforts at a healthy diet. This probability is based on the following conditional relative frequency:

$$\frac{43}{74} \approx 0.581 = 58.1\%$$

STUDENT PAGE 104

(3) d. This data set can be studied in several ways. If students examine the given categories, there is a great enough difference in the "Very Serious Effort" conditional relative frequencies to suggest an association (31% to 43%). If, however, students were to combine the "Very Serious Effort" and the "Somewhat Serious Effort" to suggest a table dividing the responses into simply "Serious" versus "Not Serious," then the conditional relative frequencies are 78% for men and 84% for women. This difference is not that large and would suggest no association. Allow students to work with this data in alternative ways to develop their answers.

4. a. The data describing 62% and 38% represent conditional information.

b. The following cells of this table can be completed as below.

The remaining cells of the table cannot be completed. Information given on either one of the marginal totals or one of the conditional frequencies for the people watching less than 3 hours a day is needed.

b. Consider a sample of 1000 Americans. Is it possible to complete the following table from the data given? Can you fill in any of the marginal or joint totals? If so, which ones? If not, explain what other information you would need in order to complete the table.

| | | Income Level | | |
		$25,000 or Less	More than $25,000	Totals
Hours Watching TV	3 or More			
	Fewer than 3			
	Totals			1000

5. In September, 1996, researchers reported at a meeting of the American Society for Microbiology that a new oral vaccine that could protect babies against a sometimes fatal diarrheal disease might be available in 1998. The disease, *rotavirus,* is the most common cause of diarrhea in the world. The vaccine was given to 1190 babies, and 1208 babies received a placebo. Within two years, researchers counted 57 episodes of rotavirus in the vaccine group and 184 in the placebo group.

a. Copy and complete the following two-way table.

	Vaccine Group	Placebo Group	Totals
Rotavirus			
No Rotavirus			
Totals			

b. What proportion of the babies in the study got rotavirus within the two years?

c. What proportion of the vaccine group got rotavirus within the two years?

d. What proportion of the placebo group got rotavirus within the two years?

e. What does it mean to say that there is no association between what group the baby was in and whether or not the baby got rotavirus?

4. a.

Income Levels

		$25,000 or Less	More Than $25,000	Totals
Hours Watching TV	3 or More	260	160	420
	Fewer than 3			580
	Totals			1000

5. a.

	Vaccine Group	Placebo Group	Totals
Rotavirus	57	184	241
No Rotavirus	1133	1024	2157
Totals	1190	1208	2398

b. $\frac{241}{2398} \approx 0.101 = 10.1\%$

c. $\frac{57}{1190} \approx 0.048 = 4.8\%$

d. $\frac{184}{1208} \approx 0.152 = 15.2\%$

e. To state that there is no association would indicate the percent of babies who got rotavirus would be approximately the same within the two groups.

f.

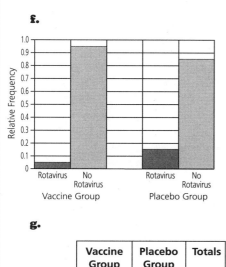

g.

	Vaccine Group	Placebo Group	Totals
Rotavirus	120	121	241
No Rotav.	1070	1087	2157
Totals	1190	1208	2398

h. The previous table is used to start the design of the simulation; however, the numbers involved are too large to duplicate the corresponding steps discussed in Lesson 11. As a result, one possible format would be to consider how the table in part g would look if the population sampled were 200 people. Consider the following representation of this sample, which uses rounded numbers. This table represents the expected frequencies with no association.

	Vaccine Group	Placebo Group	Totals
Rotavirus	1	1	2
No Rotav.	98	100	198
Totals	99	101	200

Use 200 slips of paper with 99 marked with a "V" for vaccine group and 101 marked with a "P" for placebo group. Mix the slips of paper and randomly select 2 to represent the 2 babies with rotavirus.

STUDENT PAGE 105

f. Make a bar graph showing the distribution of the conditional relative frequencies. Use a grid like the following.

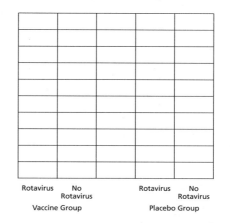

Rotavirus   No Rotavirus   Rotavirus   No Rotavirus

Vaccine Group   Placebo Group

g. Construct a two-way table of expected values, under the assumption that there is no association between the groups and whether or not the babies got rotavirus.

	Vaccine Group	Placebo Group	Totals
Rotavirus			
No Rotavirus			
Totals			

h. Describe how you would design and conduct a simulation to determine if there is an association between what group the baby was in and whether or not the baby got rotavirus.

Count the "V" slips, and record as a trial. Replace the slips and repeat this process at least 25 times. The possible outcomes are 0, 1, and 2. Determine the probability for each outcome. The experimental group would also need to be scaled to a sample of 200 to represent these observed frequencies.

	Vaccine Group	Placebo Group	Totals
Rotavirus	0.25	1.75	2
No Rotav.	98.75	99.25	198
Totals	99	101	200

Students should discuss the chance of getting a 0 outcome in their simulation if the 0.25 represents 0. The steps shoud be described, as the actual simulation need not be carried out.

STUDENT PAGE 106

6. a.

	Lived	Died	Totals
Carbolic Acid Used	34	6	40
Carbolic Acid Not Used	19	16	35
Totals	53	22	75

b.

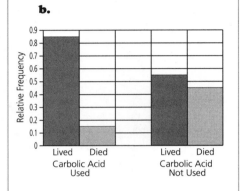

c. 53 out of 75 patients lived, for approximately 70%. If no association is assumed between the use of carbolic acid and a patient's living or dying, one would expect 70% of the 40 patients to live. The value 28 is used to complete the following table.

	Lived	Died	Totals
Acid Used	28	12	40
Acid Not Used	25	10	35
Totals	53	22	75

d. The total of 34 patients who lived is considerably greater than the expected value of 28, suggesting an association.

e. Steps should include the following: Mark 53 slips of paper with an "L" for lived and 22 with a "0" for died. Mix the slips and select 40 for the 40 patients whose rooms were disinfected. Count the slips marked "L" for the number of patients who lived if the experiment is independent of the use of carbolic acid. Return the slips and repeat this process at least 25 times. For each

6. Joseph Lister (1827–1912) was a surgeon at the Glasgow Royal Infirmary and one of the first to believe in the germ theory of infection. In one of his experiments, he used carbolic acid to disinfect operating rooms and compared the results from surgeries in these rooms to the results from rooms that had not been disinfected. Of the 40 patients in rooms where carbolic acid had been used, 34 lived. Of the 35 patients in rooms where carbolic acid had not been used, only 19 lived.

a. Construct a two-way table of observed values.

b. Make a bar graph using a grid like the following to show the distribution of the conditional relative frequencies.

Lived   Died   Lived   Died
Carbolic Acid Used   Carbolic Acid Not Used

c. Construct a two-way table of expected values.

d. Does it appear that the results of the surgeries are associated with the use of carbolic acid in the operating rooms? Explain.

e. Design and run a simulation to confirm your opinion in part c. Record the steps that you followed and the simulation results.

f. Write a paragraph explaining your conclusions on the association between the use of disinfectant and the success of surgeries.

trial, record the number of patients who lived. This list provides the data needed for estimating whether or not the 34 patients who lived survived by chance.

f. Students are expected to determine the probability that a 34 will occur in their data collection to explain whether or not carbolic acid is associated with a patient's living or dying. As the number of times a 34 occurred in the simulation is expected to be very small, association is supported by the

simulation. An association means that a connection is suggested between the use of carbolic acid and whether a patient lives or dies.

Analyzing Survey Results

Materials: list of students in school
Technology: graphing calculator

Pacing: $\frac{1}{2}$ class period to organize project and one week for students to complete the analysis

Overview

This project allows students to apply what they have learned in this module to analyze data collected from a survey of the student body. The process of organizing and conducting a survey are presented in a series of steps. The steps are designed to give students a format to help them achieve success in completing their surveys.

Teaching Notes

You may wish to have your students work in groups of two or three to complete the survey. The process of collecting data and running the simulation to analyze the data can be time-consuming. With two or three students working together, this process will take less time. Since the main idea is to decide if there is an association between two variables, it is important that students choose two variables that they think might be related. Each of the variables should have only two options for answers. To make the simulation easier to design and run, students should have only two rows and two columns in their table. They may need to consolidate some of their data to meet these constraints. Students might investigate if there is an association between the sex of the person and how the person answered a particular question. They also could investigate if there is an association between the class of a student and how the student answered a particular question.

Follow-Up

Students could give oral reports on their findings. Also, if some groups did similar questionnaires, the groups could compare and contrast their findings.

STUDENT PAGE 107

PROJECT

Analyzing Survey Results

In lesson 5, the research project conducted by students at Rufus King High School was briefly discussed. In Lessons 9 through 11, the questions from the student survey concerning diet and breakfast were analyzed. The active research process conducted by the students consisted of asking questions, collecting data, analyzing the data, summarizing conclusions, and then refining and developing more questions. Frequently, successful research raises more questions than it answers. A diagram of the research process looks like this:

OBJECTIVE

Analyze results of your own survey.

Project Analyzing Survey Results

In order to gain an insight into statistical methods, you are encouraged to carry out a random survey on a topic of interest to you.

Steps

i. Identify the population from which you will collect the information. This may be the entire student body or a particular class.

ii. Write a questionnaire that contains at least two questions. Carefully word your questions to avoid any misinterpretations or misunderstandings.

iii. Give your questionnaire to students in your group and your teacher, and have them give you feedback. Make any revisions necessary.

STUDENT PAGE 108

iv. Select a random sample of at least 30 people from the population you have identified. Recall from Lesson 3 the characteristics of random samples.

Issues that you need to address:

    a. How will you choose your random sample?

    b. How will you contact each person randomly selected?

    c. How will the surveys be returned to you?

    d. What will you do if someone does not respond to your questions?

v. Collect your results. Analyze and summarize the results both numerically by finding percents and graphically by constructing a bar graph or a similar summary.

vi. Decide on two questions from your survey that you think might be associated.

    a. Place the results from these two questions in a two-way table of observed frequencies.

    b. Complete a two-way table of expected frequencies based on the assumption that the two questions are independent.

    c. Does there appear to be an association between the two questions?

    d. Design and run a simulation to confirm your opinion about association.

    e. What are your conclusions concerning whether or not there is an association between the two questions?

vii. Write an article for your school newspaper or Web page describing your results and the limitations of your survey.

Teacher Resources

NAME _____

A student is interested in the probability that a tack will land point up when tossed. The student conducted an experiment and collected the following data. Under the Outcome column, "Up" indicates that the tack landed point up and "Side" indicates that the tack landed on its side.

Toss Number	Outcome	Cumulative Frequency That Tack Landed Point Up	Relative Frequency That Tack Landed Point Up
1	Up	_____	_____
2	Side	_____	_____
3	Side	_____	_____
4	Side	_____	_____
5	Up	_____	_____
6	Side	_____	_____
7	Side	_____	_____
8	Up	_____	_____
9	Side	_____	_____
10	Side	_____	_____
11	Side	_____	_____
12	Up	_____	_____

1. Complete the table above.

2. Construct a line graph of the relative frequency that the tack landed up.

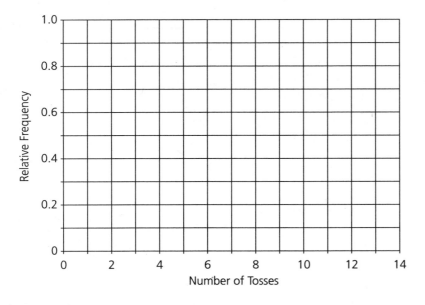

3. What is your estimate for the probability that a tack will land point up when tossed?

NAME _____

According to the U.S. Bureau of the Census, 71% of families in the United States have 0 or 1 child under 18 years of age.

1. Assume that the Census Bureau randomly selects 8 families. Design and conduct a simulation that shows the approximate distribution of the number of families that have 0 or 1 child under 18 years of age. Show all the steps of the simulation, including what random numbers you used and what constituted one trial. Then plot the resulting data. You should have at least 10 trials.

2. What is the approximate probability that out of 8 families selected, all 8 families will have 0 or 1 child under 18 years of age?

3. How many families out of 8 do you expect to have 0 or 1 child under 18 years of age?

4. Assume that the Census Bureau needs to survey 4 families with 0 or 1 child in the family. Design and conduct a simulation to determine the average number of families the Census Bureau randomly selects in order to find 4 families that meet the requirement of 0 or 1 child. Show all the steps of the simulation, including what random numbers you used and what constituted one trial. Then plot the resulting data. You should have at least 10 trials.

NAME _____

The following data summarize the teaching experience
and educational background of all the teachers in a large
high school.

		Teaching Experience	
		Fewer than 5 Years	5 Years or More
Educational Background	Less than a Master's Degree	35	18
	Master's Degree or More	12	51

1. How many teachers are in the high school?

2. How many teachers have fewer than 5 years of experience?

Let circle A represent teachers with 5 or more years of teaching
experience and let circle B represent teachers with less than a
master's degree.

3. In the Venn diagram at the right, shade the area that represents the teachers who have 5 or more years of experience.

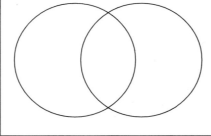

4. In the Venn diagram at the right, shade the area that represents the teachers who have 5 or more years of experience and have less than a master's degree.

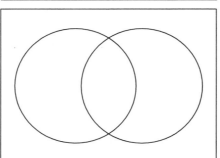

If a teacher is randomly selected, what is the probability that

5. the teacher has fewer than 5 years of experience?

6. the teacher has a master's degree or more?

7. the teacher has fewer than 5 years of experience and has a master's degree or more?

8. the teacher has fewer than 5 years of experience or has a master's degree or more?

9. the teacher does not have at least 5 years of experience?

10. the teacher has 5 or more years of experience or has 5 or fewer years of experience?

NAME _____

A random sample of 300 students were identified as male or female and then asked whether they preferred taking courses in the area of math and science or English and social studies. The results are shown in the table below.

Preferred Subject Area

	Math and Science	English and Social Studies
Males	76	52
Females	70	102

1. What is the relative frequency of students in the survey that said math and science were their preferred courses?

2. What is the relative frequency of students in the survey that are female?

3. Consider only the females in the survey. What is the conditional relative frequency of females who said that math and science were their preferred courses?

4. Consider only the males in the survey. What is the conditional relative frequency of males who said that math and science were their preferred courses?

5. Make a bar graph showing the distribution of the conditional relative frequencies.

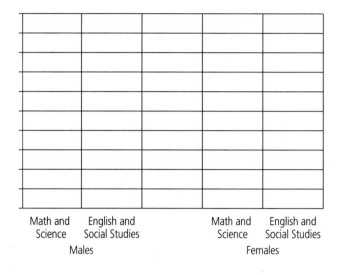

Math and   English and   Math and   English and
Science   Social Studies   Science   Social Studies
      Males   Females

6. Using the conditional relative frequencies and the bar graph, comment on whether there is an association between the sex of a person and his or her preferred courses.

7. What does it mean if there is an association between two variables?

Probability Through Data: Interpreting Results from Frequency Tables

NAME _____

Parts I and II

1. A student designed a spinner and recorded the following.

   a. Complete the table below.

Spin Number	Outcome (Area A or Area B)	Cumulative Frequency of Landing in Area A	Relative Frequency of Landing in Area A
1	A	_____	_____
2	B	_____	_____
3	A	_____	_____
4	A	_____	_____
5	B	_____	_____
6	A	_____	_____
7	B	_____	_____
8	A	_____	_____

   b. As the number of spins increases, what happens to the relative frequency that spins land in area A?

2. According to the U.S. Bureau of the Census, about 80% of the residents of the United States over the age of 25 are high-school graduates. A student designed and carried out a simulation to study the distribution of high-school graduates from 60 random samples of 5 U.S. residents over the age of 25. The results are shown below.

Number of High-School Graduates	Frequency
0	0
1	0
2	5
3	12
4	25
5	18

a. Approximate the probability of randomly selecting 5 people over the age of 25 that 4 or more would be high-school graduates.

b. What is the average number of high-school graduates from a sample of 5 people over the age of 25?

Part III

Medical records were classified by the severity of a particular condition, as well as by blood type. The results are shown in the table below.

	Blood Type			
	A	**B**	**AB**	**O**
Mild	350	50	25	50
Medium	100	50	25	100
Severe	25	13	12	50

Severity of Condition (label at left for the rows Mild, Medium, Severe)

3. If one of the records were randomly selected, what is the probability that the individual had blood type AB?

4. If one of the records were randomly selected, what is the probability that the individual had a medium condition?

5. If one of the records were randomly selected, what is the probability that the individual had blood type A and had a mild condition?

6. If one of the records were randomly selected, what is the probability that the individual had blood type A or had a mild condition?

7. If one of the records were randomly selected, what is the probability that the individual did not have type O blood?

8. Given that the record selected was a person with a mild condition, what is the probability that the individual had type A blood?

9. Given that the record selected was a person with type A blood, what is the probability that the individual had a mild condition?

Part IV

A random sample of 42 people were asked if they used a cellular phone on a regular basis, and whether or not they had had a car accident during the previous year. The results are shown in the table below.

	Had an Accident in Previous Year	Had No Accident in Previous Year
Cellular-Phone User	4	10
Non-Cellular-Phone User	8	20

10. What is the relative frequency of surveyed people that are cellular-phone users?

11. Consider only the cellular-phone users. What is the conditional relative frequency of cellular-phone users who had an accident in the previous year?

12. Consider only the non-cellular-phone users. What is the conditional relative frequency of non-cellular-phone users who had an accident in the last year?

13. Use a grid similar to the following to make a bar graph showing the distribution of conditional relative frequencies.


```
   Had an      Had No                    Had an      Had No
   Accident    Accident                  Accident    Accident
      Cellular-Phone User                   Non-Cellular-Phone User
```

14. Describe, in detail, how you would design and conduct a simulation to determine if there is an association between the two variables, *cellular-phone user* and *had an accident*.

1.

Toss Number	Outcome	Cumulative Frequency That Tack Landed Point Up	Relative Frequency That Tack Landed Point Up
1	Up	1	$\frac{1}{1} = 1.00$
2	Side	1	$\frac{1}{2} = 0.50$
3	Side	1	$\frac{1}{3} \approx 0.33$
4	Side	1	$\frac{1}{4} = 0.25$
5	Up	2	$\frac{2}{5} = 0.40$
6	Side	2	$\frac{2}{6} \approx 0.33$
7	Side	2	$\frac{2}{7} \approx 0.29$
8	Up	3	$\frac{3}{8} \approx 0.38$
9	Side	3	$\frac{3}{9} \approx 0.33$
10	Side	3	$\frac{3}{10} = 0.30$
11	Side	3	$\frac{3}{11} \approx 0.28$
12	Up	4	$\frac{4}{12} \approx 0.33$

2.

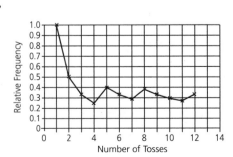

3. The line graph appears to converge to a value approximately equal to 0.3. An estimate of the probability could be in the range of 0.28 to 0.33.

1. Define a trial as selecting 8 random numbers between 1 and 100. Families having 0 or 1 child under 18 years of age would be represented by the selection of a number from 1 to 71; a selection greater than 71 would represent all other family situations. The number of families having 0 or 1 child would be simulated by the total number of values between 1 and 71 in the set of 8. This trial is repeated at least 10 times. The outcomes from the trials are used to answer the probability questions.

If students used a TI-83, the setup of a trial would be represented as **randInt(1,100,8)**.

10 or more trials would be simulated by repeated entries of this instruction.

The following example represents a simulation of 10 trials.

Trial	Random Numbers Selected	Number of Families Having 0 or 1 Child
1	67 74 12 65 73 91 10 27	5
2	2 3 29 96 35 15 89 14	6
3	99 67 90 4 91 40 17 11	5
4	62 17 8 30 76 62 21 33	7
5	56 9 65 51 5 9 91 58	7
6	50 38 94 91 15 92 54 58	5
7	74 28 42 38 88 24 49 56	6
8	96 4 59 37 37 44 6 82	6
9	39 21 47 56 98 57 32 87	6
10	15 52 66 44 55 38 36 1	8

2. Answers will vary; however, students should answer this question using the results from their simulation. Based on the selections from the above simulation, the probability that all 8 families have 0 or 1 child under 18 is $\frac{1}{10}$ = 0.10, or 10%.

3. The expected number of families having 0 or 1 child under 18 is the mean of the outcomes from the simulation. The above simulation estimates the expected outcome as 6.1. It is anticipated that most simulations will generate an expected outcome between 5 and 6. This expected outcome is also determined from the expression np in which n represents the number of selections and p is the probability of "success." For this example, the expected outcome is $(8)(0.71) = 5.68$.

4. The changes described in this problem result in the following changes in the simulation.

A trial would be represented by repeated selections of random numbers from 1 to 100 in which numbers between 1 and 71 represent the families having 0 or 1 child under 18. A trial is completed when the fourth number within that range has been obtained. The number of random numbers selected to represent the four families is the outcome of the trial. As indicated in the problem, students should develop at least 10 trials.

The following example of 10 trials was developed using random numbers generated from a calculator:

Trial	Random Numbers Selected	Number of Selections to Obtain 4 Families
1	4 54 17 76 52	5
2	83 72 20 14 85 53 24	7
3	43 1 71 100 2	5
4	90 91 82 14 65 63 79 46	8
5	50 13 70 95 63	5
6	81 64 73 32 37 50	6
7	20 43 43 48	4
8	12 34 91 5 32	5
9	96 65 26 76 95 45 21	7
10	3 19 30 98 60	5

A graph of the results follows.

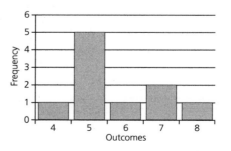

The average number of selections needed in this simulation was 5.7.

The answers to several of the questions in this quiz require the marginal totals of the following two-way table.

		Teaching Experience		
		Less than 5 Years	**5 Years or More**	**Totals**
Educational Background	**Less than a Master's Degree**	35	18	53
	Master's Degree or More	12	51	63
	Totals	47	69	116

1. 116 teachers

2. 47 teachers

3.

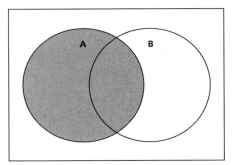

4.

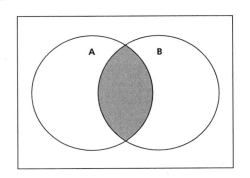

5. $\frac{47}{116} = 0.405$, or 40.5%

6. $\frac{63}{116} = 0.543$, or 54.3%

7. $\frac{12}{116} = 0.103$, or 10.3%

8. $\frac{47 + 63 - 12}{116} = 0.845$, or 84.5%

9. This answer is the same as the probability that a teacher has less than 5 years of experience,

$\frac{47}{116} = 0.405$, or 40.5%.

10. $\frac{116}{116} = 1.00$, or 100%

Several of the answers to the questions require completing the following two-way table.

Preferred Subject Area

	Math and Science	English and Social Studies	Totals
Males	76	52	128
Females	70	102	172
Totals	146	154	300

1. $\frac{146}{300} \approx 0.487 = 48.7\%$

2. $\frac{172}{300} \approx 0.573 = 57.3\%$

3. $\frac{70}{172} \approx 0.407 = 40.7\%$

4. $\frac{76}{128} \approx 0.594 = 59.4\%$

5.

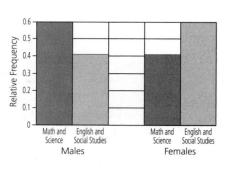

6. The conditional relative frequencies indicate a noticeable difference in the males' and females' preferred subjects. 59.4% of the males prefer math and science, while 40.7% of the females prefer these subjects. This difference in the percents is large enough to suggest an association.

7. An association between the two variables indicates that a response to one question is connected to the second question, or the selection of one item indicates a different estimate of the response to the second item. In this example, if a male were randomly selected, the estimate of his response to the preferred subject area is different from that if a female were selected. More specifically, if a male is selected, then the estimate of a preferred subject is math and science. If a female is selected, then the estimate of a preferred subject is English and social studies.

1. **a.** The following table completes the calculation of relative frequencies:

Spin Number	Outcome (Area A or Area B)	Cumulative Frequency of Landing in Area A	Relative Frequency of Landing in Area A
1	A	1	$\frac{1}{1} = 1.00$
2	B	1	$\frac{1}{2} = 0.50$
3	A	2	$\frac{2}{3} \approx 0.67$
4	A	3	$\frac{3}{4} = 0.75$
5	B	3	$\frac{3}{5} = 0.60$
6	A	4	$\frac{4}{6} \approx 0.67$
7	B	4	$\frac{4}{7} \approx 0.57$
8	A	5	$\frac{5}{8} \approx 0.63$

b. Although the number of spins recorded in this simulation are not enough to adequately identify a level of convergence discussed in the lessons, it appears the relative frequencies are approaching a value slightly more than 0.50.

2. **a.** $\frac{43}{60} \approx 0.717 = 71.7\%$

b. $\frac{2 \times 5 + 3 \times 12 + 4 \times 25 + 5 \times 18}{60} = \frac{236}{60}$

= 3.93 high-school graduates from a sample of 5 over the age of 25

3. $\frac{62}{850} \approx 0.073 = 7.3\%$

4. $\frac{275}{850} \approx 0.324 = 32.4\%$

5. $\frac{350}{850} \approx 0.412 = 41.2\%$

6. $\frac{600}{850} \approx 0.706 = 70.6\%$

7. $\frac{650}{850} \approx 0.765 = 76.5\%$

8. $\frac{350}{475} \approx 0.737 = 73.7\%$

9. $\frac{350}{475} \approx 0.737 = 73.7\%$

10. $\frac{14}{42} \approx 0.333 = 33.3\%$

11. $\frac{4}{14} \approx 0.286 = 28.6\%$

12. $\frac{8}{28} \approx 0.286 = 28.6\%$

13.

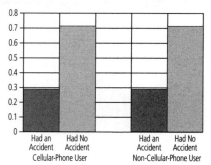

14. This simulation is modeled after the problems in Unit IV. A simulation should include the following steps:

i. Create 42 slips of paper with 12 labeled "A" for had an accident and 30 slips labeled "NA" for had no accident.

ii. Thoroughly mix the slips of paper. 14 are selected representing the 14 people with a cellular phone. Of the 14 slips selected, count the slips labeled "A." Record this value.

iii. Repeat the above process at least 25 times. Determine from the collected results the probability of observing 4 (or 4 people who had an accident and were cellular-phone users). If this probability is less than 10%, then consider the observed results as an estimate of an association of the two items. Clearly, the results from this sample are not associated or independent.

Students are expected only to describe the steps of the simulation in this problem.

Lesson 1, Problem 2

NAME _____

Class Results from 30 Tosses of a Coin

Number of Heads	Tally	Frequency	Number of Heads	Tally	Frequency
1	————	————	16	————	————
2	————	————	17	————	————
3	————	————	18	————	————
4	————	————	19	————	————
5	————	————	20	————	————
6	————	————	21	————	————
7	————	————	22	————	————
8	————	————	23	————	————
9	————	————	24	————	————
10	————	————	25	————	————
11	————	————	26	————	————
12	————	————	27	————	————
13	————	————	28	————	————
14	————	————	29	————	————
15	————	————	30	————	————

Lesson 1, Problem 7

NAME _____

Relative Frequency of Heads and Tails

Toss Number	Outcome	Cumulative Number of Heads	Relative Frequency of Heads	Cumulative Number of Tails	Relative Frequency of Tails
1	_____	_____	_____	_____	_____
2	_____	_____	_____	_____	_____
3	_____	_____	_____	_____	_____
4	_____	_____	_____	_____	_____
5	_____	_____	_____	_____	_____
6	_____	_____	_____	_____	_____
7	_____	_____	_____	_____	_____
8	_____	_____	_____	_____	_____
9	_____	_____	_____	_____	_____
10	_____	_____	_____	_____	_____
11	_____	_____	_____	_____	_____
12	_____	_____	_____	_____	_____
13	_____	_____	_____	_____	_____
14	_____	_____	_____	_____	_____
15	_____	_____	_____	_____	_____
16	_____	_____	_____	_____	_____
17	_____	_____	_____	_____	_____
18	_____	_____	_____	_____	_____
19	_____	_____	_____	_____	_____
20	_____	_____	_____	_____	_____
21	_____	_____	_____	_____	_____
22	_____	_____	_____	_____	_____
23	_____	_____	_____	_____	_____
24	_____	_____	_____	_____	_____
25	_____	_____	_____	_____	_____
26	_____	_____	_____	_____	_____
27	_____	_____	_____	_____	_____
28	_____	_____	_____	_____	_____
29	_____	_____	_____	_____	_____
30	_____	_____	_____	_____	_____

Lesson 1, Problems 11 and 14

NAME _____

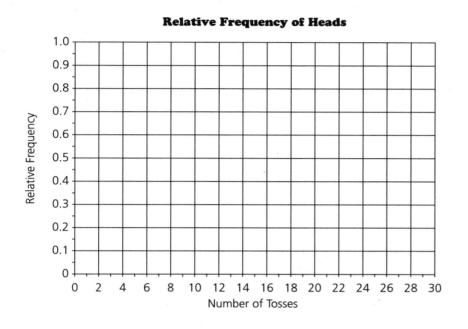

Relative Frequency of Heads

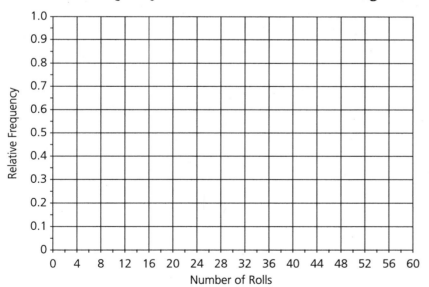

Relative Frequency of an Even Number When Rolling a Die

Lesson 2, Problem 1

NAME _____

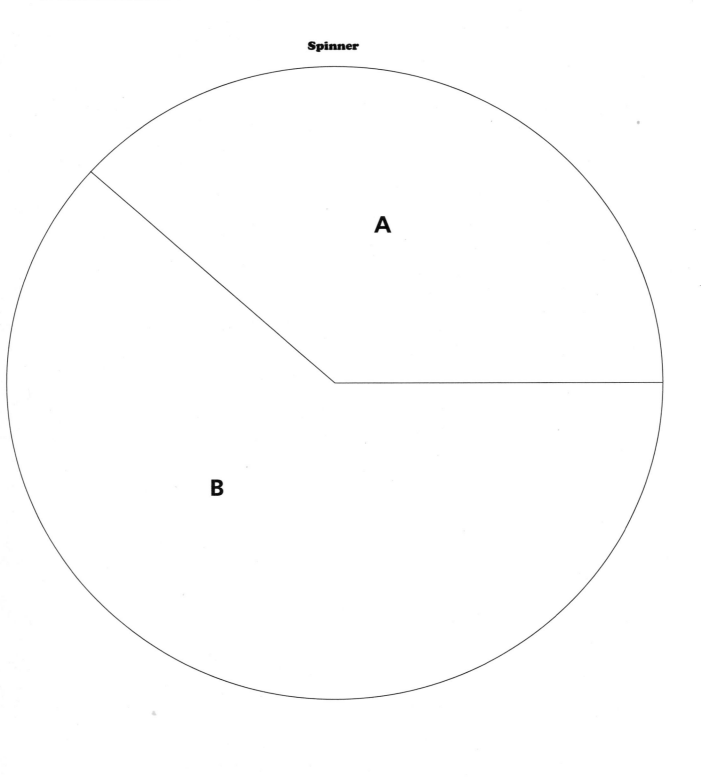

Spinner

Lesson 2, Problem 1

NAME _____

Spinner Table

Spin Number	Outcome (Area A or Area B)	Cumulative Number in Area A	Cumulative Number in Area B	Relative Frequency of Spins Landing in Area A
1	————	————	————	————
2	————	————	————	————
3	————	————	————	————
4	————	————	————	————
5	————	————	————	————
6	————	————	————	————
7	————	————	————	————
8	————	————	————	————
9	————	————	————	————
10	————	————	————	————
11	————	————	————	————
12	————	————	————	————
13	————	————	————	————
14	————	————	————	————
15	————	————	————	————
16	————	————	————	————
17	————	————	————	————
18	————	————	————	————
19	————	————	————	————
20	————	————	————	————
21	————	————	————	————
22	————	————	————	————
23	————	————	————	————
24	————	————	————	————
25	————	————	————	————
26	————	————	————	————
27	————	————	————	————
28	————	————	————	————
29	————	————	————	————
30	————	————	————	————

Lesson 2, Problems 1 and 5

NAME _____

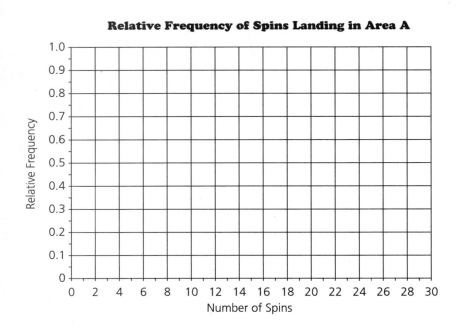

Relative Frequency of Spins Landing in Area A

y-axis: Relative Frequency (0 to 1.0)
x-axis: Number of Spins (0 to 30)

Relative Frequency of Purple Skittles

y-axis: Relative Frequency (0 to 1.0)
x-axis: Number of Purple Skittles (0 to 30)

Lesson 2, Problem 3

NAME _____

Class Results on the Number of Spins Landing in Area A

Number in Area A	Tally	Frequency	Number in in Area B	Tally	Frequency
1	————	————	16	————	————
2	————	————	17	————	————
3	————	————	18	————	————
4	————	————	19	————	————
5	————	————	20	————	————
6	————	————	21	————	————
7	————	————	22	————	————
8	————	————	23	————	————
9	————	————	24	————	————
10	————	————	25	————	————
11	————	————	26	————	————
12	————	————	27	————	————
13	————	————	28	————	————
14	————	————	29	————	————
15	————	————	30	————	————

Lesson 2, Problem 7

NAME _____

Square Pyramid

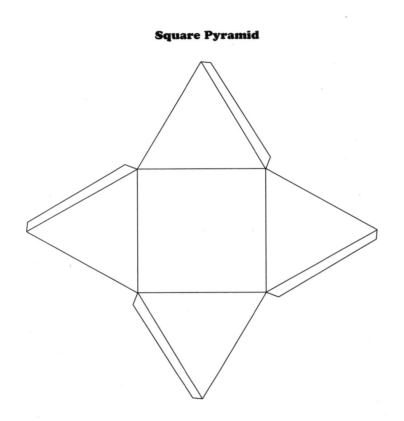

Assessment for Units I and II, Problem 1

NAME _____

M&M Number	Outcome	Cumulative Number of Blue M&Ms	Relative Frequency of Blue M&Ms
1	————	————	————
2	————	————	————
3	————	————	————
4	————	————	————
5	————	————	————
6	————	————	————
7	————	————	————
8	————	————	————
9	————	————	————
10	————	————	————
11	————	————	————
12	————	————	————
13	————	————	————
14	————	————	————
15	————	————	————

Assessment for Units I and II, Problem 1

NAME _____

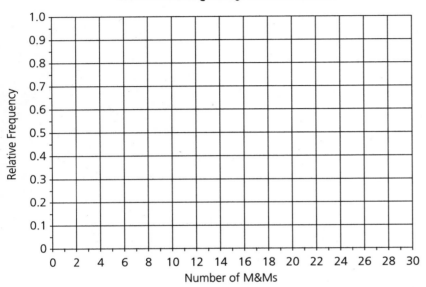

Relative Frequency of Blue M&Ms

Lesson 5, Problem 1

NAME _____

Class Results for Survey on "Do You Eat Breakfast?"

Student	Outcome (Yes or No)	Cumulative Number of Yeses	Relative Frequency of Yeses
1	_____	_____	_____
2	_____	_____	_____
3	_____	_____	_____
4	_____	_____	_____
5	_____	_____	_____
6	_____	_____	_____
7	_____	_____	_____
8	_____	_____	_____
9	_____	_____	_____
10	_____	_____	_____
11	_____	_____	_____
12	_____	_____	_____
13	_____	_____	_____
14	_____	_____	_____
15	_____	_____	_____
16	_____	_____	_____
17	_____	_____	_____
18	_____	_____	_____
19	_____	_____	_____
20	_____	_____	_____
21	_____	_____	_____
22	_____	_____	_____
23	_____	_____	_____
24	_____	_____	_____
25	_____	_____	_____
26	_____	_____	_____
27	_____	_____	_____
28	_____	_____	_____
29	_____	_____	_____
30	_____	_____	_____

Lesson 5, Problem 2

NAME _____

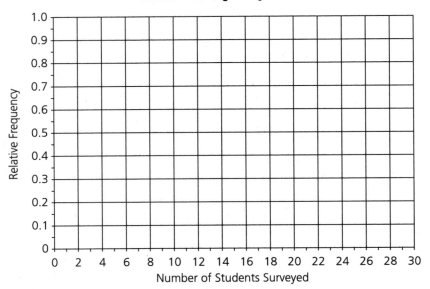

Relative Frequency of Yeses

Lesson 6, Problems 5, 6, 7, 15, and 16

NAME _____

Venn Diagrams

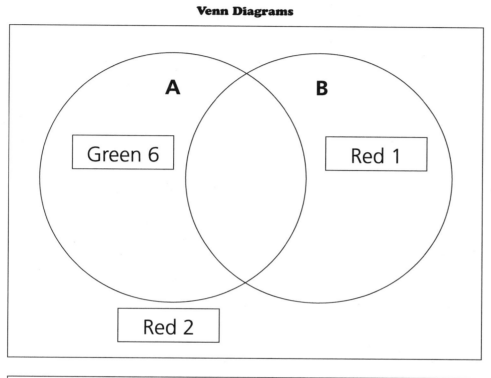

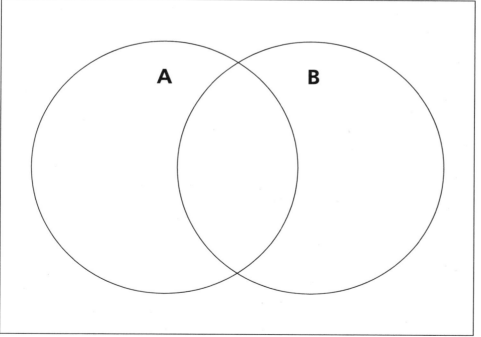